CONTEMPORARY MATHEMATICS

Titles in this Series

Rational Constructions of Modules for Simple Lie Algebras

CONTEMPORARY MATHEMATICS

Volume 5

Rational Constructions of Modules for Simple Lie Algebras

George B. Seligman

AMERICAN MATHEMATICAL SOCIETY
Providence • Rhode Island

Research supported by National Science Foundation
Grants MCS 76-10237 and MCS 79-04473.

Parts of Chapter 1 first appeared in *Representations of isotropic simple Lie algebras over general non-modular fields,* Lie Theories and their Applications, Queen's Papers in Pure Appl. Math., no. 48, Queen's University, Kingston, Ont., 1978, pp. 528–574. Reprinted with permission of the publisher.

Parts of Chapters 3 and 4 first appeared in *Mappings into symmetric powers,* J. Algebra **62** (1980), 455–472, © Academic Press, 1980. Reprinted with permission of the publisher.

Library of Congress Cataloging in Publication Data

Seligman, George B., 1927–

Rational constructions of modules for simple Lie algebras.

(Contemporary mathematics; v. 5)

Bibliography: p.

1. Lie algebras. 2. Modules (Algebra) I. Title. II. Series: Contemporary mathematics (American Mathematical Society); v. 5.

QA252.3.S437 512′.55 81-12781

ISBN 0-8218-5008-3 AACR2

ISSN 0271-4132

1980 *Mathematics Subject Classification.* Primary 17B10;
Secondary 15A24, 15A72, 16A28, 17C40.

TABLE OF CONTENTS

FOREWORD

Given a simple Lie algebra over an arbitrary field of characteristic zero, what are its finite-dimensional irreducible representations? As with most mathematical questions, this one must be sharpened to bring it down somewhere between the banal and the impossible. One could be content with answering it by taking a splitting field K for the Lie algebra L, observing that the classical theory of the highest weight gives all irreducible L_K-modules and that the irreducible L-modules all occur as L-submodules of these. On the other hand, one could insist on the specification of a finite-dimensional vector space and an action of L in every case. It is not certain whether even the classical theories of Cartan, Weyl, and Harish-Chandra in the complex case would be satisfactory to someone thoroughly committed to this strict latter view.

For the present work, it is axiomatic that the results of these classical theories are satisfactory, and they should serve as a model for a theory for non-split algebras. Starting with Cartan [5], mathematicians have been successful in refining the approach of descending from L_K-modules by Galois-theoretical methods, culminating in the elegant work of Tits [7]. (Although the context there is algebraic groups, the translation to Lie algebras in characteristic zero presents no difficulties.) The modules one is seeking are found by splitting extension of the base field, introduction of an action of the Galois group in certain modules for the split algebra, and then by the rather "implicit" selection of the fixed subspace.

Following the lines of the structure theory in [26], it is sought here to avoid splitting extensions in the realization of the representations, obtaining them by processes that are rational over the given ground field. In the present work, two essential restrictions are imposed: The first, that the simple Lie algebra be isotropic, allows a first crude parametrization of irreducible modules by highest weights relative to a maximal split torus. In the real case, this amounts to studying the real-irreducible representations of the Lie algebras of non-compact simple Lie groups. The second restriction, that the relative root-system be reduced, is rather arbitrarily imposed to keep the monograph of reasonable length. Among the real simple Lie algebras satisfying the former restriction, it excludes most of those arising as Lie algebras of (complex or quaternionic) unitary groups, as well as one class of exceptional Lie algebras of each of the dimensions 52 and 78 (of respective relative types BC_1 and BC_2 - see Propositions ℓ and n of §V.8 of [26]). Both restrictions are provisional, and indeed much of the work toward elimination of the second one has been done.

Subject to these assumptions, the program is to ascertain which highest (relative) weights can occur, and to give a rational procedure for constructing modules of given highest weight. Now and then techniques of base field extension are used to eliminate candidates for the status of highest weight, or to show that a set of modules, rationally obtained, is exhaustive for a given highest weight. It is worth emphasizing that, in a sense we make more precise below, the modules are always rationally constructed; it is only in proof that these are all the modules that we sometimes use extended coefficients.

The sense of rational constructibility that we adopt is limited to the realization of our modules by one, or a combination of, the following processes:

1) Forming quotients of specific modules for the universal enveloping algebra $U(L)$ by a submodule with given generators --- this is completely analogous to the constructions in the split case.

2) Taking submodules of the tensor product of finite-dimensional irreducible modules, usually for a fixed reductive subalgebra L_0 of L.

3) Taking irreducible modules for finite-dimensional semisimple associative algebras rationally given or constructed from data giving the structure of L.

A quick idea of the nature of the results can be obtained by skimming through Chapter I up to the definition of λ-admissible L_0-modules in §1.4, then glancing ahead to §V.7, where there are described the steps 1) - 3) appropriate to derived algebras of central simple associative algebras and to skew elements of such algebras with respect to certain involutions of first kind. Corresponding answers for other algebras are to be found in the sections headed "Summary" in subsequent chapters.

In order to reduce the "fundamental modules" to the lists in these "summaries", the methods of this work often require treatment of modules for a somewhat larger set of highest weights, then a reduction of these to the smaller list. Chapters III and IV deal with questions in the theory of associative algebras that must be answered to achieve the reduction, as well as to construct the more complicated of the fundamental modules. They may be of independent interest for their treatment of presentations for the symmetric powers of an associative algebra and for the structure of such powers when the algebra is related to the Lie algebras of Chapters V and VI. It has been my good fortune to have as colleagues David Saltman and Tsuneo Tamagawa, who have been very helpful concerning the structure of symmetric powers.

The use of symmetric powers emerges as the counterpart for non-split algebras of relative type A to the tensor representations of GL(n) or SL(n) in the split case, as presented in Chapter IV of The Classical Groups, Hermann Weyl's definitive treatise on that subject. That reference is abundantly, and necessarily, cited in our Chapter IV. The symmetric powers also provide us with the means to complete the construction of modules for algebras of relative type C, in Chapters V and VI.

From our point of view, the representations of the Lie algebras of orthogonal groups must be approached along different lines. To some extent, this is inherent in the methods; moreover, the spin - or half-spin representations necessitate going beyond tensors and their analogues. Here (in Chapter VII) it is these spin-representations that hold the key; the fundamental tensor representations occur as modules for the associated meson algebra, realized within the tensor square of the Clifford algebra.

A persuasive argument for the methods exposed here is to be found in the cases of the exceptional algebras. When the simple algebra L is one of the larger exceptional algebras (say a form of E_7 or E_8), the subalgebra L_0 whose modules give those for L via the methods of induced modules is usually (up to center) one of the smaller exceptional algebras (say a form of G_2 or F_4). Moreover, the L_0-modules needed to yield a fundamental set of L-modules turn out to be just the most familiar ones (Chapters VIII-XI.)

To come to terms with the context of the problem and the presentation, one should be familiar with large parts of the books on Lie algebras by Jacobson [15] and by Humphreys [11], particularly with those parts dealing with representations of split semi-simple algebras, and to some extent with my [26] for terminology, conventions and results in the non-split case. It may be well to have at hand Weyl's Classical Groups [28], the treatise of Curtis and Reiner [7], and Jacobson's book on Jordan algebras [17] in connection with Chapters IV-VII. The chapters on exceptional algebras refer to a more specialized bibliography, and are somewhat less self-contained than the rest of the work.

Portions of the work have been read by David Saltman and by Gregg Zuckerman, to both of whom I am grateful. Kevin Carlin has been a most valuable reader of the proofs. Mary Ellen DelVecchio, who prepared the camera-ready copy, and Cathy Belton, who typed the original manuscript, have shown patience, cheerfulness and pride in their work while making my task easier. Finally, I am grateful to Academic Press, Inc., publishers of the Journal of Algebra, for permission to reprint material that makes up much of Chapter III.

CHAPTER I

GENERALITIES ON FINITE-DIMENSIONAL MODULES

The ground field F is taken to be of characteristic zero, and L will denote a semisimple Lie algebra over F. We denote by T a maximal split torus in L, i.e., a commutative subalgebra such that ad h is diagonalizable in F for every $h \in T$, and which is maximal with this property. We assume $T \neq 0$, an assumption which for $F = \mathbb{R}$ amounts to assuming that L is not compact. The objective of this chapter will be to develop in general terms the analogy between finite-dimensional irreducible L-modules, analyzed through their restrictions to T, and the classical theory when L is split, i.e., when the maximal split torus T is a full Cartan subalgebra of L, as in Chapter VII of [15].

1. BASIC OBSERVATIONS ON IRREDUCIBLE MODULES.

For the decomposition of L relative to T, details are to be found in [26], Chapter 1. The essential structural facts for us are the following:

There is a decomposition of L into root spaces L_α ($\alpha \neq 0$ in T^*) and the centralizer L_0 of T, where $L_\beta = \{x \mid [xh] = \beta(h)x$ for all $h \in T\}$. Thus

$$L = L_0 + \sum_\alpha L_\alpha .$$

Those $\alpha \neq 0$ with $L_\alpha \neq 0$ form a system of roots (in the general sense of Bourbaki [3]) in their rational span, or in the real space obtained by extending the scalars of this rational space [26]. This space may be (linearly) ordered, in such a way that $N = \sum_{\alpha > 0} L_\alpha$ is a subalgebra that acts nilpotently on every finite-dimensional L-module, while the action of T on such a module is always diagonalizable in F. (Here one appeals to the theory of modules for the split three-dimensional simple Lie algebra, as developed in Chapter III of [15]).

If we let $\alpha_1, \ldots, \alpha_r$ be a fundamental system of (positive) roots, for example the set of those positive roots which cannot be written as sums of positive roots, then L is generated, as Lie algebra, by L_0 and the $L_{\pm\alpha_i}$, $1 \leq i \leq r$. If L is simple, as will be the case in most of this work, L

is generated by the $L_{\pm\alpha_i}$, and $L_0 = \sum_i [L_{\alpha_i} L_{-\alpha_i}]$.

Now T normalizes N, and N acts nilpotently: thus any finite-dimensional (right) L-module V will contain $v \neq 0$ with $vN = 0$, and the set of such v is T-stable. Therefore, for certain elements $\lambda \in T^*$ there are non-zero elements $v \in V$ with $vN = 0$, $vh = \lambda(h)v$ for all $h \in T$. Each such λ is called a highest weight of V, and for each such λ the space of all $v \in V$ satisfying the conditions is called the space of λ-extreme vectors of V. Since L_0 normalizes N and centralizes T, the space of λ-extreme vectors is an L_0-submodule.

PROPOSITION I.1. *Let V be an irreducible L-module. Then:*

a) *If λ is a highest weight of V, the space W of λ-extreme vectors in V is an irreducible L_0-module.*

b) *V is linearly generated over F by all*

$$w\, e_{\beta_1} \cdots e_{\beta_s}, \quad s \geq 0; \quad e_\beta \in L_\beta, \ \beta < 0; \quad w \in W,$$

and by all

$$w\, e_{-\alpha_{i_1}} \cdots e_{-\alpha_{i_s}}, \quad s \geq 0, \quad e_{-\alpha_i} \in L_{-\alpha_i}; \quad w \in W.$$

c) *W is the set of elements $w \in V$ with $wN = 0$.*

d) *$W = \{v \in V \mid vh = \lambda(h)v$ for all $h \in T\}$.*

Conversely, if V is a finite-dimensional L-module, generated by $W = \{v \in V \mid wN = 0\}$, and if W is an irreducible L_0-module, then V is irreducible.

PROOF. In the case of b), the fact that L_0 normalizes each L_α assures that each of the sets of vectors described has an L_0-stable span. Each of these spans is clearly stable under the action of all $L_{-\alpha_i}$; to see that they are stable under all L_{α_i}, and hence under L, use induction on s and the following facts: $wL_{\alpha_i} = 0$;
$[L_{-\alpha_j} L_{\alpha_i}] = 0$ if $i \neq j$; $[L_{-\alpha_i} L_{\alpha_i}] \subseteq L_0$; $[L_\beta L_{\alpha_i}] \subseteq L_{\beta + \alpha_i}$, where either $L_{\beta+\alpha_i} = 0$, or $\beta + \alpha_i$ is a negative root, if $\beta < 0$, $-\beta$ is not fundamental. Thus each of the spans of b) is a submodule, so must be V.

Evidently the reasoning of the last paragraph applies when W is replaced by any nonzero L_0-submodule. But $\alpha_1, \ldots, \alpha_r \in T^*$ are known to be linearly independent, and each $(we_{-\alpha_{i_1}} \cdots e_{-\alpha_{i_s}})h =$
$(\lambda - \alpha_{i_1} \cdots \alpha_{i_s})(h)\, we_{-\alpha_{i_1}} \cdots e_{-\alpha_{i_s}}$ for all $h \in T$, $w \in W$. It follows that the only elements of W which are linear combinations of the second set of vectors prescribed in b) are those obtained as combinations of those for which

$s = 0$. The irreducibility of W follows.

To see that c) and d) hold, note that since $V^N = \{v \in V \mid vN = 0\}$ is T-stable, the only way that c) can fail is to have $\mu \neq \lambda$, $\mu \in T^*$, and $0 \neq u \in V^N$ with $uh = \mu(h)u$ for all $h \in T$. But then from b), $\mu = \lambda - \alpha_{i_1}, \ldots, \alpha_{i_s}$, $s > 0$, and V is generated by $S = \{v \mid v \in V^N, vh = \mu(h)v, \text{ all } h \in T\}$ exactly as by W in b). By comparison of weights, the submodule generated by S cannot contain W, so c) holds. From b) and our remarks in connection with a) we have d).

The converse follows from the complete reducibility of V (see, for instance, Chapter III of [15].) Writing $V = V_1 \oplus \ldots \oplus V_n$, where the V_i are irreducible, the nilpotency of N in each V_i shows that $W \cap V_i \neq 0$ if $V_i \neq 0$, so that $W \cap V_i$ is a nonzero L_0-submodule of W. Thus $W \subseteq V_i$ and $V_i = V$.

PROPOSITION I.2. Let V_i, $i = 1,2$, be irreducible L-modules, $W_i = V_i^N$ as above. Let $\varphi: W_1 \to W_2$ be an isomorphism of L_0-modules. Then there is a unique isomorphism $\eta: V_1 \to V_2$ of L-modules extending φ.

PROOF. One forms the product $V_1 \times V_2$, and lets V be the L-submodule generated by the graph W of φ. Evidently V is spanned by all elements

$$(we_{-\alpha_{i_1}} \ldots e_{-\alpha_{i_s}}, \varphi(w)\, e_{-\alpha_{i_1}} \ldots e_{-\alpha_{i_s}}), \quad w \in W_1,$$

projects onto V_1 resp. V_2 in the respective projections (as in Prop. 1, b)), and meets $W_1 \times W_2$ in the graph of φ, by comparison of weights. The last implies that V is a proper L-submodule of $V_1 \times V_2$, and the preceding that it is neither V_1 nor V_2. It follows that V_1 and V_2 are isomorphic, the projections π_1, π_2 from V to V_1 resp. V_2 being isomorphisms. The composite $V_1 \xrightarrow{\pi_1^{-1}} V \xrightarrow{\pi_2} V_2$ is the desired isomorphism η. It is clearly unique.

Now suppose L is simple, and let Z be the centroid of L, i.e., the centralizer of ad L in $\mathrm{Hom}(L,L)$. Then Z is a finite field extension of F (see Chapter X of [15]); the action of Z on L (written on the left) makes L into a Z-vector space, and the same bracket operation on L makes this Z-space into a Lie algebra over Z, denoted by $L(Z)$. Then $L(Z)$ is simple and has Z as centroid, so remains simple upon extension of the base field. The next observation is that every non-trivial irreducible L-module is an $L(Z)$-module.

PROPOSITION I.3. Let V be an irreducible L-module, $\dim_F V > 1$. Assume that L is simple with centroid Z. Then V carries a structure of Z-vector space $V(Z)$, relative to which the original action of L makes $V(Z)$ into an irreducible $L(Z)$-module.

PROOF. Form the L-module $V \otimes_F L$, as usual: $(v \otimes x)y = vy \otimes x + v \otimes [xy]$ for $v \in V$; $x,y \in L$. Then for $\zeta \in Z$, the mapping $1 \otimes \zeta$ sends $V \otimes_F L$ into itself, and this action of Z on $V \otimes_F L$ makes $V \otimes_F L$ into a Z-vector space. The actions of Z and of L evidently commute, so that $V \otimes_F L$ is actually an $L(Z)$-module.

By assumption, $VL \neq 0$, where VL is the image of $V \otimes_F L$ under the homomorphism of L-modules $V \otimes_F L \to V$ sending $v \otimes x$ to vx. Since V is irreducible, $VL = V$, and $V \otimes_F L$ has V as homomorphic image under a homomorphism of L-modules. By the complete reducibility of all L-modules, $V \otimes_F L$ contains an L-submodule isomorphic to V; call one such V_0.

Also, $L(Z)$ is semisimple over Z, so the $L(Z)$ module $V \otimes_F L$ is completely reducible. Thus $V \otimes_F L$ is the direct sum of irreducible $L(Z)$-submodules $U_1, \ldots, U_s$, with $U_iL = U_i$, and possibly also of $U_0 = \{u \in V \otimes_F L \mid uL = 0\}$. Thus $V_0 = V_0L \subseteq U_1L + \ldots + U_s L \subseteq U_1 \oplus \ldots \oplus U_s$.

Now each U_i, $1 \leq i \leq s$, is an irreducible L-module; for if M is a nonzero L-submodule of U_i, ML is also an L-submodule, and cannot be zero because the subset of U_i annihilated by $L = L(Z)$ is a submodule different from U_i. In fact, ML is an $L(Z)$-module, since for $m \in M$, $x \in L$, $\zeta \in Z$,

$$\zeta(mx) = m(\zeta x) \in ML .$$

By the irreducibility of the $L(Z)$-module U_i, $ML = U_i$. But $ML \subseteq M \subseteq U_i$, so $M = U_i$, and U_i is irreducible as L-module. It follows that V_0, hence V, is isomorphic as L-module to some U_i, $1 \leq i \leq s$.

Each U_i is a Z-vector space and an $L(Z)$-module. Thus if $\eta: U_i \to V$ is a homomorphism of L-modules, we may define $\zeta\eta(u_i) = \eta(\zeta u_i)$ for all $u_i \in U_i$, $\zeta \in Z$, and transport thereby the structure of Z-vector space from U_i to V. This action of Z on V evidently centralizes that of L, so V is thus an $L(Z)$-module, necessarily irreducible. This proves Proposition 3.

2. RELATIONS IN IRREDUCIBLE L-MODULES.

Let L be simple over F as above, and let V be a finite-dimensional irreducible L-module with $VL \neq 0$. By Proposition 3, we may assume L is central simple, i.e., that the centroid Z of L is the ground field F.

Let W be as in Proposition 1, with λ and the $\{\alpha_i\}$ also as in that proposition. Now T has a basis $h_1, \ldots, h_r$, where h_i is an element of $T \cap [L_{\alpha_i} L_{-\alpha_i}]$ such that $\alpha_i(h_i) = 2$, and is unique if one requires that $h_i = [e_{\alpha_i} e_{-\alpha_i}]$ for some $e_{\pm\alpha_i} \in L_{\pm\alpha_i}$ ([26], p. 6). Whenever

$0 \neq e_{\alpha_i} \in L_{\alpha_i}$, there is an $e_{-\alpha_i} \in L_{-\alpha_i}$ such that $[e_{\alpha_i} e_{-\alpha_i}] = h_i$ ([26],pp. 4-6); thus one can apply the representation theory of the split three-dimensional Lie algebra generated by $e_{\pm\alpha_i}$ to see that $\lambda(h_i)$ _is a non-negative integer_ for each i. By analogy with the split case, we refer to such $\lambda \in T^*$ as _dominant integral functions_.

If V, W, λ, $\{\alpha_i\}$, $\{h_i\}$ are as above, we fix an index i. Let $0 \neq f \in L_{-\alpha_i}$. Then, as above there is $e \in L_{\alpha_i}$ with $[ef] = h_i$. If $0 \neq w \in W$, the theory of the three-dimensional algebra applies to $< e,f,h_i >$, to show that w, wf, $wf^2 = (wf)f$, ... span an irreducible submodule for $< e,f,h_i >$ of V. We have $wf^{m_i} \neq 0$, $wf^{m_i+1} = 0$, where $\lambda(h_i) = m_i \in \mathbb{Z}^+$.

PROPOSITION I.4. _Let_ V, W, λ, $\{\alpha_i\}$, $\{h_i\}$ _be as above, and fix an index_ i. _Then for every_ $v \in W$, _and for all_ $f_1, \ldots, f_{m+1} \in L_{-\alpha_i}$,

$$vf_1 \cdots f_{m+1} = 0,$$

where $m = \lambda(h_i)$.

PROOF. First suppose $2\alpha_i$ is not a root, and let $f_1, \ldots, f_{m+1}$ be as above. Then $[f_i f_j] = 0$ for all i,j, and if $\xi_1, \ldots, \xi_{m+1}$ are arbitrary in F, we have

$$v\left(\sum_{i=1}^{m+1} \xi_i f_i\right)^{m+1} = 0 \tag{1}$$

by the remarks just preceding the proposition. Upon expansion, we find the coefficient of the product $\xi_1 \cdots \xi_{m+1}$ to be $(m+1)!\ vf_1 \cdots f_{m+1}$. This coefficient must be zero, because the relation (1) holds identically in $\xi_1, \ldots, \xi_{m+1}$ from the infinite field F. This completes the proof when $2\alpha_i$ is not a root.

Now suppose $2\alpha_i$ is a root, so that $\pm\alpha_i$, $\pm 2\alpha_i$ are the roots (F-) linearly dependent on α_i. Replacing α_i by $2\alpha_i$, we find a unique $h \in T$ of the form $h = [xy]$, $x \in L_{2\alpha_i}$, $y \in L_{-2\alpha_i}$, with $(2\alpha_i)(h) = 2$, or $\alpha_i(h) = 1$. If β is any root, we have $(2\alpha_i, 2\alpha_i)\,\beta(h) = 2(\beta, 2\alpha_i)$ and $(\alpha_i, \alpha_i)\,\beta(h_i) = 2(\beta, \alpha_i)$, where the inner products in T^* are induced by duality from the restriction to T of the Killing form of L ([26,§§ I.1,2). Since $(\alpha_i, \alpha_i) \neq 0$, we have $2h = h_i$. As above, $\lambda(h) = n$ is a non-negative integer with $vy_1 \cdots y_{n+1} = 0$ for all $y_1, \ldots, y_{n+1} \in L_{-2\alpha_i}$. From $h_i = 2h$, we see that $m = 2n = \lambda(h_i)$, so $\lambda(h_i)$ is an _even_ integer if $2\alpha_i$ is a root.

Now we prove by induction on k, starting with $k = 0$, the following assertion: _Let_ $0 \leq k \leq n$; _let_ $f_1, \ldots, f_{2k+1} \in L_{-\alpha_i}$; _let_ $y_1, \ldots, y_{n-k} \in L_{-2\alpha_i}$. _Then_

$$vy_1 \cdots y_{n-k} f_1 \cdots f_{2k+1} = 0. \tag{2}$$

The assertion (2) for $k = n$ is Proposition 4.

For $k = 0$, we first show $vy^n f = 0$ for all $y \in L_{-2\alpha_i}$, $f \in L_{-\alpha_i}$, and we may assume $y \neq 0$. (If $n = 0$, we have $\lambda(h_i) = 0$ and $vf = 0$ by the remarks preceding the proposition.) Then $[y\, L_{\alpha_i}] = L_{-\alpha_i}$; this is seen by noting that $L_0 + L_{\alpha_i} + L_{-\alpha_i} + L_{2\alpha_i} + L_{-2\alpha_i}$ is a module for $< x,y,h >$ as above, with $L_{\alpha_i} + L_{-\alpha_i}$ a submodule in which L_{α_i} is the subspace where h has the eigenvalue 1, $L_{-\alpha_i}$ that for -1; then the claim follows from the structure of $< x,y,h >$-modules. Now we have $vy^{n+1} = 0$, so that $vy^{n+1}e = 0$, where $e \in L_{\alpha_i}$, $[ye] = f$. Since f commutes with y and $ve = 0$, we have

$$0 = vy^{n+1}e = \sum_{j=0}^{n+1} vy^j\, f\, y^{n-j} = (n+2)\, vy^n f.$$

Thus $vy^n f = 0$.

An argument using polarization, as in the case where $2\alpha_i$ is not a root, now applies to show

$$vy_1 \cdots y_n f = 0$$

for all $y_1, \ldots, y_n \in L_{-2\alpha_i}$, $f \in L_{-\alpha_i}$. The case $k = 0$ is done.

Now let $k > 0$, and assume our assertion proved for $k-1$. First let y,f,e be as in the case $k = 0$, so that $0 = vy^{n-k+1} f^{2k-1}$ by inductive hypothesis, and therefore $0 = vy^{n-k+1} f^{2k}e$. Now

$$vy^{n-k+1} f^{2k}e = vy^{n-k+1} f^{2k-1}e\, f + vy^{n-k+1} f^{2k-1} [fe]$$

$$= vy^{n-k+1} f^{2k-2}\, e\, f^2 + vy^{n-k+1} f^{2k-2} [fe]f,$$

since the last term of the previous line is zero by induction. The final term here is

$$vy^{n-k+1} f^{2k-1} [fe] + vy^{n-k+1} f^{2k-2} [[fe]f],$$

of which the first term is zero by hypothesis, as well as the second, which has the form $vy^{n-k+1} f_1^{2k-2} f_2$, f_1 and f_2 in $L_{-\alpha_i}$. That is, we have $vy^{n-k+1} f^{2k-2}\, e\, f^2 = 0$. The argument of this paragraph may now be repeated so as to arrive at $vy^{n-k+1}\, e\, f^{2k} = 0$.

Now, as for $k = 0$, $vy^{n-k+1}\, e\, f^{2k}$ is equal to $\sum_{j=0}^{n-k+1} vy^j\, f\, y^{n-k-j} f^{2k} = (n-k+2)\, vy^{n-k} f^{2k+1}$, so this quantity is zero. Since $k \leq n$, $vy^{n-k} f^{2k+1} = 0$. Polarization on y again yields

$$vy_1 \cdots y_{n-k} f^{2k+1} = 0 \tag{3}$$

for all $y_1, \ldots, y_{n-k} \in L_{-2\alpha_i}$, $f \in L_{-\alpha_i}$.

Finally we polarize f; let $f_1, \ldots, f_{2k+1} \in L_{-\alpha_i}$, $\xi_1, \ldots, \xi_{2k+1} \in F$. Then for all $y_1, \ldots, y_{n-k}$, (3) yields

(4) $$vy_1 \cdots y_{n-k} \left(\sum_{j=1}^{2k+1} \xi_j f_j \right)^{2k+1} = 0.$$

The coefficient of $\xi_1 \cdots \xi_{2k+1}$ in (4) is the sum, over all permutations π of the indices, of the terms

(5) $$vy_1 \cdots y_{n-k} f_{\pi(1)} \cdots f_{\pi(2k+1)} .$$

If $\pi(j) > \pi(j+1)$, the term (5) is equal to

$$vy_1 \cdots y_{n-k} f_{\pi(1)} \cdots f_{\pi(j+1)} f_{\pi(j)} \cdots f_{\pi(2k+1)}$$
$$+ vy_1 \cdots y_{n-k} f_{\pi(1)} \cdots [f_{\pi(j)} f_{\pi(j+1)}] \cdots f_{\pi(2k+1)}$$

The second term here is of the form

$$vy_1 \cdots y_{n-k} y_{n-k+1} f_1' \cdots f_{2k-1}' ,$$

so is zero by induction. It follows that each of the terms (5) is equal to $vy_1 \cdots y_{n-k} f_1 \cdots f_{2k+1}$, so that our coefficient of $\xi_1 \cdots \xi_{2k+1}$ is

$$(2k+1)! \; vy_1 \cdots y_{n-k} f_1 \cdots f_{2k+1} ,$$

which must then be zero as before. The proof by induction is now complete, and Proposition 4 is proved.

When a basis for L is taken to consist of a basis for L_0, combined with the bases of root-vectors for $N = \sum_{\alpha > 0} L_\alpha$ and for $N^- = \sum_{\alpha < 0} L_\alpha$, we order the basis so that the part in N comes first, then the basis for L_0, then that for N^-. From the theorem of Poincare-Birkhoff-Witt (PBW) there results a direct decomposition of the universal enveloping algebra $U(L)$:

(6) $$U(L) = U(L_0) \oplus NU(N + L_0) \oplus U(L_0 + N^-) N^- \oplus NU(L)N^- .$$

We denote by ψ the projection of $U(L)$ onto $U(L_0)$ determined by this decomposition.

Now let the setting be that of Proposition 4, and let $e_1, \ldots, e_{m+1}$ be arbitrary elements of L_{α_i}. By Proposition 4, W is annihilated by

$$u = f_1 \cdots f_{m+1} e_{m+1} \cdots e_1 \in U(L).$$

In $U(L)$, we have for $h \in T$, $[uh] = 0$; meanwhile $NU(N + L_0)$ has a basis consisting of elements c with $[ch] = \sum_{m_j \in \mathbb{Z}^+} m_j \alpha_j(h)c$ for all $h \in T$, some $m_j > 0$; $U(L_0 + N^-)N^-$ has a basis of such elements c with $[ch] = -\sum_{m_j \in \mathbb{Z}^+} m_j \alpha_j(h)c$ for all h, some $m_j > 0$; $[U(L_0),h] = 0$, all

$h \in T$; and $NU(L)N^-$ has a basis of common eigenvectors for all $\mathrm{ad}\, h$, $h \in T$. It follows that

$$u \in U(L_0) + NU(L)N^-$$

Now $WN = 0$, so $\psi(u)$ must annihilate W. This completes the proof of

PROPOSITION I.5. *Let the notations be as in Proposition 4. Let* $e_1, \ldots, e_{m+1}$ *be arbitrary in* L_{α_i}, *and let* ψ *be the projection of* $U(L)$ *on* $U(L_0)$ *determined by the decomposition* (6). *Then* $\psi(f_1 \ldots f_{m+1}\, e_{m+1} \ldots e_1)$ *annihilates* W.

3. INDUCED MODULES. CONSTRUCTION OF IRREDUCIBLE MODULES.

Let $P = L_0 + N$, a ("minimal F-parabolic") subalgebra of L, and let W be an irreducible (right) L_0-module. Then W becomes a P-module if we define $WN = 0$, and, accordingly, a right $U(P)$-module, where $U(P)$ is the universal enveloping algebra, a subalgebra of $U(L)$. As left $U(P)$-module, $U(L)$ is free (by PBW), with basis 1 and the non-empty monomials in a basis for N^- consisting of root-vectors for negative roots. Thus $U(P)$ is a direct summand of $U(L)$ as left $U(P)$-module. (In (6), $U(P)$ is the sum of the first two terms, and the last two give our complementary summand.) As in [11], § 20, the tensor product

$$W' = W \otimes_{U(P)} U(L)$$

is a right $U(L)$-module by right multiplication of $U(L)$ on the second factor, and thus an L-module. It is the *induced module*, sometimes denoted

$$W' = \mathrm{ind}_{U(P)}^{U(L)} W.$$

The mapping $w \mapsto w \otimes 1$ maps W, as $U(P)$-module, isomorphically onto a subspace of W', which we hereby identify with W.

It is clear that the $U(L)$-module W' and the inclusion mapping $W \hookrightarrow W'$ are characterized by the following properties:

i) W' is an L-module containing W as P-submodule, and W' is generated by W.

ii) If V is an L-module, $\varphi \in \mathrm{Hom}_{U(P)}(W,V)$, then there is a unique homomorphism η of L-modules such that the diagram

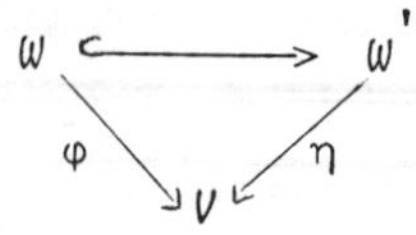

is commutative.

It should be emphasized that no assumptions are made in the above concerning the dimensions of W, W', V. We say that the irreducible P-module W has weight λ if $\lambda \in T^*$ and if for all $h \in T$, $w \in W$, $wh = \lambda(h)w$. In the sequel we assume there is $\lambda \in T^*$ such that W has weight λ.

When W has weight λ, and b runs over the non-empty ordered monomials in a basis for N^- consisting of root-vectors, we have for each such b, and for all $h \in T$,

$$[bh] = -\sum_{j=1}^{r} n_j \alpha_j(h)b,$$

for non-negative integers n_j, not all zero, so that $(w \otimes b)h = wh \otimes b + w \otimes [bh] = (\lambda - \Sigma_{j=1}^{r} n_j \alpha_j)(h)(w \otimes b)$ for b as above and for all $w \in W$, $h \in T$. From the fact that these b, together with 1, form a basis for the free left $U(P)$-module $U(L)$, we see that W is the subspace of W' belonging to the weight λ.

It follows that if M_1 and M_2 are $U(L)$-submodules of W' with $M_i \cap W = 0$, $i = 1,2$, then $(M_1 + M_2) \cap W = 0$. For both M_1 and M_2 are sums of weight spaces for T, and the weight λ cannot occur in either, therefore cannot occur in their sum. We now have the following proposition, which contains another proof of Proposition 2:

PROPOSITION I.6. Let L, T, $\{L_\alpha\}$ be as in §1, and let P be as above. Let W be an irreducible L_0-module having weight λ. Then there is a unique irreducible L-module V with highest weight λ and such that $W \cong V_\lambda$ as L_0-module. It is obtained by taking the quotient of $W' = W \otimes_{U(P)} U(L)$ by its unique maximal submodule M.

PROOF. If a submodule of W' intersects W non-trivially, it contains W by L_0-irreducibility, and therefore is all of W'. By the remarks immediately preceding the proposition, it follows that W' has a unique maximal $(\neq W')$ submodule M. Then the irreducible L-module W'/M is a sum of weight-spaces for weights of the form $\lambda - \Sigma n_j \alpha_j$, $n_j \geq 0$ in $\mathbb{Z}$, with only the cosets of elements of W having weight λ. That is, if $V = W'/M$ and if V_λ is the subspace of weight λ, the canonical map $W \to V$ is an L_0-isomorphism of W onto V_λ.

As in Proposition 1 it follows that V_λ is the subspace of V annihilated by N. In this sense, V (now possibly of infinite dimension) is an irreducible L-module of highest weight λ. The uniqueness of V follows by the characteristic property ii) of W' and the uniqueness of M.

Necessary conditions for V to be finite-dimensional were developed in §2: 1) $\lambda(h_i)$ must be a non-negative integer for each i, and must be even if $2\alpha_i$ is a root; 2) For each i, if $\lambda(h_i) = m_i \geq 0$, and if $w \in W$, then for all $f_1, \ldots, f_{m_i+1} \in L_{-\alpha_i}$,

$$(w \otimes 1)f_1 \ldots f_{m_i+1} = w \otimes f_1 \ldots f_{m_i+1} \in M .$$

That is, <u>the set of all</u> $w \otimes f_1 \ldots f_{m_i+1}$, $w \in W$, $1 \leq i \leq r$, $f_1, \ldots, f_{m_i+1} \in L_{-\alpha_i}$, <u>must generate a proper submodule of</u> W'. We next show these conditions are <u>sufficient</u>:

<u>PROPOSITION I.7</u>. <u>Let W be a finite-dimensional irreducible L_0-module having weight λ. Assume that $\lambda(h_i)$ is a non-negative integer for $1 \leq i \leq r$, with $\lambda(h_i)$ even if $2\alpha_i$ is a root. Then there is a finite-dimensional L-module V with highest weight-space L_0-isomorphic to W if and only if the L-submodule of $W' = W \otimes_{U(P)} U(L)$ generated by all $w \otimes f_1 \ldots f_{\lambda(h_i)+1}$, as in 2) above, is proper.</u>

<u>PROOF</u>. If the submodule R generated as above is proper, it fails to meet W, except in zero. A maximal submodule M with respect to containing R and not meeting W is again a maximal submodule, as in the proof of Proposition 6. Thus what must be shown is that <u>if</u> R is proper, then M is of finite codimension in W'. Our argument for this is only a mild modification of that for the split case to be found in [11], §21.2.

We work in V, regarding W as L_0-submodule of V. Then for $w \in W$, we have: $wh = \lambda(h)w$, all $h \in T$; $wx = 0$ if $x \in L_{\alpha_i}$, $1 \leq i \leq r$; $wf_1 \ldots f_{\lambda(h_i)+1} = 0$ if $f_1, \ldots, f_{\lambda(h_i)+1} \in L_{-\alpha_i}$, $1 \leq i \leq r$. Moreover, V is generated by W. We show these conditions imply the finite-dimensionality of V.

First we fix i, $1 \leq i \leq r$, and let $L_i = L_0 + L_{\alpha_i} + L_{-\alpha_i} + L_{2\alpha_i} + L_{-2\alpha_i}$, a subalgebra of L, and consider the subspace of V spanned by all $wf_1 \ldots f_t$ for $w \in W$, $f_i \in L_{-\alpha_i}$, $t < \lambda(h_i) + 1$. This is a finite-dimensional subspace and an L_i-submodule containing W; to see it is L_0-stable, note that $[L_0, L_{-\alpha_i}] \subseteq L_{-\alpha_i}$ and that W is an L_0-submodule; it is trivially $L_{-\alpha_i}$-stable; that it is L_{α_i}-stable follows from $[L_{-\alpha_i} L_{\alpha_i}] \subseteq L_0$ and $WL_{\alpha_i} = 0$. That our space is an L_i-submodule now follows because L_0 and $L_{\pm\alpha_i}$ generate L_i as in [26], §I.2.

Still with i fixed, let V' be the sum of all finite-dimensional L_i-submodules. We have $W \subseteq V'$ by the last paragraph, and if X is any finite-dimensional L_i-submodule, we have $XL \subseteq X + \Sigma_\beta XL_\beta$, where β runs over all roots $\neq \pm\alpha_i$, $\pm 2\alpha_i$. If β is such a root, XL_β is finite-dimensional, L_0-stable, and for $\gamma = \pm\alpha_i$, $XL_\beta L_\gamma \subseteq XL_\beta + XL_{\beta+\gamma}$. Thus XL is an L_i-submodule, necessarily finite-dimensional, so is contained in V'. It follows that $V'L \subseteq V'$, and, since $W \subseteq V'$ and generates V, that

$V' = V$. Thus, for each i, V is a sum of finite-dimensional L_i-submodules.

In each finite-dimensional L_i-submodule, all elements of $L_{\pm\alpha_i}$ and of $L_{\pm 2\alpha_i}$ act nilpotently. Thus the action on V of each $x \in L_{\alpha_i} \cup L_{-\alpha_i}$ is locally nilpotent, and this for each i. In particular, if i is fixed, and if $e \in L_{\alpha_i}$, $f \in L_{-\alpha_i}$ are chosen with $[ef] \in T$, $\alpha_i([ef]) = 2$, the F-endomorphism

$$A = \exp(f)\exp(e)\exp(f)$$

of V is well-defined by the (locally finite) power series.

As in §1, V is a sum of weight-spaces for T, each of the weight-spaces is finite-dimensional, and all weights have the form $\lambda - \Sigma_j n_j \alpha_j$, $n_j \in \mathbb{Z}^+$. When i is fixed, we therefore have that any weight-space V_μ is contained in a finite-dimensional L_i-submodule. Forming the operator A as above for this i, A stabilizes our L_i-submodule containing V_μ. Now if $v \in V_\mu$, $h \in T$,

$$\begin{aligned}(vA)h &= (((vA)h)A^{-1})A \\ &= (v(h \exp(-\mathrm{ad}\, f)\exp(-\mathrm{ad}\, e)\exp(-\mathrm{ad}\, f)))A = (v(h^{\sigma_i}))A,\end{aligned}$$

where $h^{\sigma_i} = h$ if $\alpha_i(h) = 0$, $h_i^{\sigma_i} = -h_i$. Thus

$$v(h^{\sigma_i}) = (\mu(h) - \mu(h_i)\,\alpha_i(h))v = \mu^{\sigma_i}(h)v,$$

where in this last σ_i denotes the Weyl reflection $\mu \to \mu - \mu(h_i)\alpha_i$, acting in T^* (cf. [26], pp. 26-7). That is, A maps V_μ into $V_{\mu^{\sigma_i}}$; applying the same argument to $V_{\mu^{\sigma_i}}$, we see that A maps this into V_μ. But A is invertible on finite-dimensional L_i-modules, so maps V_μ onto $V_{\mu^{\sigma_i}}$.

Now the (relative) Weyl group is the group of automorphisms of T^* generated by the σ_i, $1 \le i \le r$. It follows from the last paragraph that the weights of V are permuted by the Weyl group, and weight-spaces for weights conjugate under the Weyl group have the same dimension.

To show that V has finite dimension one now shows that the set of weights is finite, exactly as in §21.2 of [11]: For given a weight μ, we have $\mu = \lambda - \Sigma\, n_j \alpha_j$ as above, and we may replace μ by a conjugate to assume $\Sigma\, n_j$ is minimal for the orbit of μ. Since $\mu^{\sigma_i} = \mu - \mu(h_i)\alpha_i$ is in the orbit of μ, and since $\mu(h_i) \in \mathbb{Z}$, we must have $\mu(h_i) \ge 0$ for all i; that is, μ is dominant.

Now $h_1, \ldots, h_r$ are a basis for T; $\alpha_1, \ldots, \alpha_r$ are a basis for T^*; and all $\lambda(h_i)$ are integers. It follows that λ is a rational combination of $\alpha_1, \ldots, \alpha_r$, as is μ. But if a rational combination of $\alpha_1, \ldots, \alpha_r$

is _dominant, then all coefficients are non-negative_. This follows from the positive-definiteness of the inner product on the rational span of the $\{\alpha_j\}$ and from the fact that $\alpha_j(h_i)$ is a positive multiple of (α_j, α_i) (cf. [26], §I.1). Thus if μ is dominant, and if $\mu = \Sigma\, m_i\, \alpha_i - \Sigma\, n_j\, \alpha_j$, where α_k are distinct with all m_i, $n_j \in \mathbb{Z}^+$, all $n_j > 0$, we have
$(\mu + \Sigma\, n_j\, \alpha_j, \Sigma\, n_j\, \alpha_j) \geq 0$, > 0 unless $\Sigma\, n_j\, \alpha_j$ is vacuous, while
$(\mu + \Sigma\, n_j\, \alpha_j, \Sigma\, n_j\, \alpha_j) = (\Sigma\, m_i\, \alpha_i, \Sigma\, n_j\, \alpha_j) \leq 0$ since $(\alpha_i,\alpha_j) \leq 0$ of $i \neq j$.
It follows that $(\mu + \Sigma\, n_j\, \alpha_j, \Sigma\, n_j\, \alpha_j) = 0$, and that $\Sigma\, n_j\, \alpha_j$ is indeed vacuous, proving our last underlined assertion. Recalling that μ has the form $\lambda - \Sigma\, k_j\, \alpha_j$, $k_j \in \mathbb{Z}^+$, we see that there are only finitely many possibilities for μ. This completes the proof of Proposition 7.

We can be more precise about the maximal submodule M, consisting of those relations that must be introduced in W' to yield our V. Namely, our assumption was that the submodule R of W' generated by all $wf_1 \ldots f_{\lambda(h_i)+1}$ was distinct from W'. We claim $R = M$, more precisely:

PROPOSITION I.8. _Let the hypotheses on_ W _be as in Proposition_ 7, _and let_ R _be the_ L-_submodule of_ $W \otimes_{U(P)} U(L) = W'$ _generated by all_ $w \otimes f_1 \ldots f_{\lambda(h_i)+1}$ _as in Proposition_ 7. _If_ $R \neq W'$, _then_ R _is maximal in_ W', _and_ W'/R _is the unique finite-dimensional irreducible_ L-_module with highest weight space_ L_0-_isomorphic to_ W.

PROOF. Reviewing the proof of Proposition 7, now working in W'/R, we see that the sum of all finite-dimensional L_i-submodules of W'/R is L-stable and contains W for each i. Since W still generates W'/R, we again have that W'/R is the sum of its finite-dimensional L_i-submodules. The rest of the proof of Proposition 7 applies equally well to W'/R, which is therefore _finite-dimensional_. Thus W'/R is a completely reducible L-module, in which the weight-space for λ is W. It follows from the irreducibility of W as L_0-module that λ can only occur as a weight in one irreducible summand of W'/R; but then this summand contains W, which generates W'/R. That is, W'/R is irreducible, and Proposition 8 is proved.

4. λ-ADMISSIBLE L_0-MODULES.

A finite-dimensional irreducible L_0-module W will be called λ-admissible if W has weight λ, which is dominant integral, and if the submodule R of $W' = W \otimes_{U(P)} U(L)$ as defined in the last section is distinct from W'. From §§2,3 we see that this amounts to requiring that W be L_0-isomorphic to the highest weight space of a finite-dimensional irreducible L-module of highest weight λ. In this section we develop some tests for recognizing λ-admissible L_0-modules (L simple).

The notations of §3 continue. Evidently $\mathcal{R} \neq \mathcal{W}'$ is equivalent to $\mathcal{R} \cap \mathcal{W} = 0$.

<u>LEMMA I.1. A necessary and sufficient condition that $\mathcal{R} \cap \mathcal{W} \neq 0$ is that for some $w \in \mathcal{W}$ and some sequences</u>

$$f_1, \ldots, f_{\lambda(h_i)+1}, \text{ from } L_{-\alpha_i};$$
$$e_1, \ldots, e_{\lambda(h_i)+1}, \text{ from } L_{\alpha_i},$$

<u>we have</u> $w \otimes f_1 \ldots f_{\lambda(h_i)+1} e_{\lambda(h_i)+1} \ldots e_1 \neq 0$.

<u>PROOF</u>. Since the displayed element in $\mathcal{R}$ belongs to the weight λ, and since $\mathcal{W}'_\lambda = \mathcal{W}$, the condition is sufficient. For necessity, suppose $\mathcal{W} \cap \mathcal{R} \neq 0$. Since L is simple, L is generated by the $L_{\pm\alpha_j}$; we see by comparing weights that there must be a non-zero element of the form

$$v = w \otimes f_1 \ldots f_{\lambda(h_i)+1} x_1 \ldots x_s, \tag{7}$$

belonging to the weight λ. Here α_i is fixed, $f_k \in L_{-\alpha_i}$, and $x_1, \ldots, x_s \in \bigcup_{j=1}^{r} (L_{\alpha_j} \cup L_{-\alpha_j})$. For this fixed i, we assume w; $f_1, \ldots, f_{\lambda(h_i)+1}$; $x_1, \ldots, x_s$ such that s is minimal with $v \neq 0$. Clearly $s \geq \lambda(h_i)+1$. We show there must be elements $e_1, \ldots, e_{m+1} (m = \lambda(h_i))$ among $x_1, \ldots, x_s$, such that $e_1, \ldots, e_{m+1} \in L_{\alpha_i}$ and

$$v = w \otimes f_1 \ldots f_{m+1} e_{m+1} \ldots e_1 y_1 \ldots y_{s-(m+1)}$$

for some $y_1, \ldots, y_{s-(m+1)}$ (hence $s = m+1$, by minimality). This yields the lemma.

Let j, $1 \leq j \leq s$, be the first index in (7) such that x_j belongs to a positive root, say $x_j \in L_{\alpha_k}$. If $j > 1$, we can move x_j to the left past all $x_\nu \notin L_{-\alpha_k}$ until perhaps encountering $x_\nu \in L_{-\alpha_k}$. Then $x_\nu x_j = x_j x_\nu + [x_\nu x_j]$, the second term here in L_0. Since L_0 normalizes all $L_{\pm\alpha_j}$ and is contained in P, we obtain that moving x_j to the left of x_ν changes (7) by the addition of a term in $\mathcal{W}'_\lambda$ of the form

$$w' \otimes f'_1 \ldots f'_{m+1} y_1 \ldots y_{s-2},$$

which must be zero by the minimality of s. Thus (7) is equal to $w \otimes f_1 \ldots f_{m+1} x_j x_1 \ldots \hat{x}_j \ldots x_s$, and we may assume $x_1 \in L_{\alpha_k}$ in (7). If $k \neq i$, x_1 commutes with all f_ν, is in P, and annihilates w, so that (7) cannot be different from zero. Therefore $k = i$, and we have $x_1 = e_{m+1} \in L_{\alpha_i}$. Now suppose we have shown

$$v = w \otimes f_1 \ldots f_{m+1} e_{m+1} \ldots e_{r+1} y_1 \ldots y_t,$$

where $1 \leq r \leq m$, $t + (m-r) = s$, and where $(y_1, \ldots, y_t, e_{r+1}, \ldots, e_{m+1})$ is a permutation of $(x_1, \ldots, x_s)$.

By the minimality of s, the first y_j to lie in some L_{α_k} may again be assumed to be y_1. (There must be such y_j, since otherwise $v \notin W'_\lambda$.) Now if $\alpha_k \neq \alpha_i$, we have $0 \neq w \otimes f_1 \ldots f_{m+1} e_{m+1} \ldots e_{r+1} y_1 \in W'_{\lambda - r\alpha_i + \alpha_k}$; but $\lambda - r\alpha_i + \alpha_k$ is not a weight of W', so this is absurd. Therefore $y_1 \in L_{\alpha_j}$; we define $e_r = y_1$, and have

$$v = w \otimes f_1 \ldots f_{m+1} e_{m+1} \ldots e_r z_1 \ldots z_{t-1},$$

satisfying our conditions with r replaced by $r-1$. This carries the induction downward on r to $r = 0$, where

$$0 \neq w \otimes f_1 \ldots f_{m+1} e_{m+1} \ldots e_1 y_1 \ldots y_{s-(m+1)},$$

as desired. The lemma is proved.

With W a finite-dimensional irreducible λ-admissible L_0-module, $m = \lambda(h_i)$, we have seen in Proposition 5 that if $f_1, \ldots, f_{m+1} \in L_{-\alpha_i}$; $e_1, \ldots, e_{m+1} \in L_{\alpha_i}$, then $\psi(f_1 \ldots f_{m+1} e_{m+1} \ldots e_1) \in U(L_0)$ annihilates W. Thus if $\lambda \in T^*$ is dominant integral, and if we define $J(\lambda)$ to be the ideal in $U(L_0)$ generated by all

$$\psi(f_1 \ldots f_{m_i+1} e_{m_i+1} \ldots e_1), \quad m_i = \lambda(h_i),$$

$$f_1, \ldots, f_{m_i+1} \in L_{-\alpha_i}; \; e_1, \ldots, e_{m_i+1} \in L_{\alpha_i}; \quad 1 \leq i \leq r,$$

then $J(\lambda)$ annihilates each irreducible λ-admissible L_0-module.

On the other hand, the definition of the mapping ψ shows that any irreducible module W for L_0 having weight λ and annihilated by $J(\lambda)$ has the property that for all $w \in W$, all i; all $f_1, \ldots, f_{m_i+1} \in L_{-\alpha_i}$; all $e_1, \ldots, e_{m_i+1} \in L_{\alpha_i}$,

$$\begin{aligned} w \otimes f_1 \ldots f_{m_i+1} e_{m_i+1} \ldots e_1 &= w \otimes \psi(f_1 \ldots e_1) \\ &= w\,\psi(f_1 \ldots e_1) \otimes 1 = 0. \end{aligned}$$

Now Lemma 1 applies to give $W \cap R = 0$, so that W is λ-admissible. That is, we have proved

THEOREM I.1. *Let L be a central simple Lie algebra over F, $T \neq 0$ a maximal F-split torus in L, L_0 the centralizer of T, $\{\alpha_1, \ldots, \alpha_r\}$ a fundamental system of roots relative to T. Let $\lambda \in T^*$ be a dominant integral function relative to $\{\alpha_1, \ldots, \alpha_r\}$, such that $\lambda(h_i)$ is even if $2\alpha_i$ is a root. Let W be an irreducible L_0-module on which each $h \in T$ acts as the scalar $\lambda(h)$. Then W occurs as the highest weight space of a* finite-dimensional irreducible L-module if and only if the annihilator of W in $U(L_0)$ contains the ideal $J(\lambda)$ defined above.

The next lemma is the essential step in showing that $J(\lambda)$ is of finite codimension in $U(L_0)$:

LEMMA I.2. Let $U(L_0)$ be filtered by powers of its generating set L_0, so that $U(L_0)_m$ is the subspace spanned by products of at most m factors from L_0. Let $e \in L_{\alpha_i}$, $f \in L_{-\alpha_i}$. Then for each positive integer j there are positive integers $a_{j0}, a_{j1}, \ldots, a_{jj}$ such that $f^j e^j \in U(L)$ is congruent, modulo $\sum_{k=0}^{j-1} L_\alpha^k\, U(L_0)_{j-k-1}\, L_{-\alpha}^k$, to $\sum_{k=0}^{j} a_{jk} e^k [fe]^{j-k} f^k$.

PROOF. For $j = 1$, we have $fe = [fe] + ef$, so $a_{10} = a_{11} = 1$, and the Lemma holds. We show that its validity for $j(\geq 1)$ implies its validity for $j + 1$.

First it will be noted that from $L_0\, L_\alpha \subseteq L_\alpha L_0 + L_\alpha$ we have $U(L_0)_m\, L_\alpha \subseteq L_\alpha U(L_0)_m$ for each m, and likewise $L_{-\alpha} U(L_0)_m \subseteq U(L_0)_m\, L_{-\alpha}$. Now we show that, for $0 \leq k \leq j-1$,

$$f\, L_\alpha^k\, U(L_0)_{j-1-k}\, L_{-\alpha}^k e \subseteq \sum_{r=0}^{j} L_\alpha^r\, U(L_0)_{j-r}\, L_{-\alpha}^r = S. \tag{8}$$

For, if $f_1, \ldots, f_k \in L_{-\alpha}$, then

$$f_1 \ldots f_k e = e f_1 \ldots f_k + \Sigma_{\nu=1}^k f_1 \ldots f_{\nu-1} z_\nu f_{\nu+1} \ldots f_k,$$

where $z_\nu = [f_\nu e] \in L_0$. By the remarks above, $f_1 \ldots f_{\nu-1} z_\nu f_{\nu+1} \ldots f_k \in U(L_0)_1\, L_{-\alpha}^{k-1}$. Thus the left-hand side of (8) is contained in $f\, L_\alpha^k\, U(L_0)_{j-1-k} e\, L_{-\alpha}^k + f\, L_\alpha^k\, U(L_0)_{j-k}\, L_{-\alpha}^{k-1}$. The remarks cited show that the first member is contained in $f\, L_\alpha^{k+1}\, U(L_0)_{j-1-k}\, L_{-\alpha}^k$.

A procedure analogous to the above gives $f\, L_\alpha^s \subseteq L_\alpha^s f + L_\alpha^{s-1}\, U(L_0)_1$, so that

$$\begin{aligned} f\, L_\alpha^{k+1}\, U(L_0)_{j-1-k}\, L_{-\alpha}^k &\subseteq L_\alpha^{k+1}\, f U(L_0)_{j-1-k}\, L_{-\alpha}^k \\ &\quad + L_\alpha^k\, U(L_0)_{j-k}\, L_{-\alpha}^k \\ &\subseteq L_\alpha^{k+1}\, U(L_0)_{j-1-k}\, L_{-\alpha}^{k+1} + L_\alpha^k U(L_0)_{j-k}\, L_{-\alpha}^k \subseteq S, \end{aligned}$$

and likewise $f\, L_\alpha^k\, U(L_0)_{j-k}\, L_{-\alpha}^{k-1} \subseteq S$. Thus the inductive step required on $f^{j+1}e^{j+1} = f(f^j e^j)e$ reduces to the consideration, modulo S, of all elements

$$fe^k [fe]^{j-k} f^k e, \quad 0 \leq k \leq j. \tag{9}$$

Now $f^k e = ef^k + \Sigma_{\nu=0}^{k-1} f^\nu [fe] f^{k-\nu-1} \equiv ef^k + k[fe]f^{k-1}$ (modulo $L_{-\alpha}^{k-1}$), and likewise $fe^k \equiv e^k f + ke^{k-1}[fe]$ (modulo L_α^{k-1}). Therefore (9) is equal, modulo S, to

$$(10)\qquad e^k f[fe]^{j-k} ef^k + ke^k f[fe]^{j+1-k} f^{k-1}$$
$$+ ke^{k-1}[fe]^{j+1-k} ef^k + k^2 e^{k-1}[fe]^{j+2-k} f^{k-1}$$
$$+ [\text{terms in } e^k f\, U(L_0)_{j-k} L_{-\alpha}^{k-1} + e^{k-1} U(L_0)_{j+1-k} L_{-\alpha}^{k-1}$$
$$+ L_\alpha^{k-1} U(L_0)_{j-k} ef^k + L_\alpha^{k-1} U(L_0)_{j+1-k} f^{k-1}] .$$

The second and fourth members inside the square brackets are obviously in S, and the first and third of this set also belong to S by $L_{-\alpha}U(L_0)_{j-k} \subseteq U(L_0)_{j-k} L_{-\alpha}$ and $U(L_0)_{j-k} L_\alpha \subseteq L_\alpha U(L_0)_{j-k}$. Thus we need only consider the first four terms of (10), of which the fourth has the form desired in the statement of the Lemma.

From $f[fe]^{j+1-k} \equiv [fe]^{j+1-k} f$ (modulo $U(L_0)_{j-k} L_{-\alpha}$) and $[fe]^{j+1-k} e \equiv e\,[fe]^{j+1-k}$ (modulo $L_\alpha U(L_0)_{j-k}$), both the second and third terms are congruent to $ke^k[fe]^{j+1-k}f^k$ (modulo S). The first term is $e^k[fe]^{j-k}fef^k$ (modulo $e^k U(L_0)_{j-k-1} L_{-\alpha} ef^k$), and this module is contained in $L_\alpha^k U(L_0)_{j-k-1}\, e\, L_{-\alpha}^{k+1} + L_\alpha^k U(L_0)_{j-k} L_{-\alpha}^k \subseteq S$. Furthermore, $e^k[fe]^{j-k}f\,e\,f^k = e^k[fe]^{j+1-k} f^k + e^k[fe]^{j-k} ef^{k+1}$, and $e^k[fe]^{j-k}\, e\, f^{k+1} \equiv e^{k+1}[fe]^{j-k} f^{k+1}$ (modulo S) by an argument like that earlier in this paragraph. It follows that, modulo S, (10), and therefore (9), is congruent to

$$e^{k+1}[fe]^{j+1-(k+1)} f^{k+1} + (2k+1)\, e^k[fe]^{j+1-k} f^k$$
$$+ k^2 e^{k-1} [fe]^{j+1-(k-1)} f^{k-1} ,$$

the last term occurring only if $k \geq 1$. Now our coefficients $a_{j+1,k}$ are non-negative integers: the coefficient $a_{j+1,0}$ of $[fe]^{j+1}$ is $a_{j,0} + a_{j,1} > 0$, we have $a_{j+1,j+1} = a_{jj} > 0$ (in fact, $a_{jj} = 1$), $a_{j+1,j} = a_{j,j-1} + (2j+1)a_{jj}$, and, for $1 \leq k \leq j$,

$$a_{j+1,k} = a_{j,k-1} + (2k+1)a_{j,k} + k^2 a_{j,k+1} .$$

PROPOSITION I.10. <u>Let λ be dominant integral,</u> $m = \max\ \{\lambda(h_i),\ 1 \leq i \leq r\}$. <u>Then $U(L_0)/J(\lambda)$ has dimension at most equal to</u> $(m+1)^{[L_0:F]}$. <u>Only the irreducible modules for this algebra can occur as highest weight-spaces for finite-dimensional irreducible L-modules of highest weight λ. In particular, the number of λ-admissible irreducible L_0-modules is finite.</u>

<u>PROOF</u>. Since L is assumed simple, $L_0 = \sum_{i=1}^{r} [L_{\alpha_i}\ L_{-\alpha_i}]$. Thus L_0 has a

basis of elements $[ef]$, where $e \in L_{\alpha_i}$, $f \in L_{-\alpha_i}$ and where i takes on (perhaps) all values $1, \ldots, r$. If $m_i = \lambda(h_i)$, Lemma 2 gives

$$f^{m_i+1} e^{m_i+1} \equiv \sum_{k=0}^{m_i+1} a_{m_i+1,k}\, e^k [fe]^{m_i+1-k} f^k$$

$$\left(\text{modulo} \sum_{k \le m_i} L_\alpha^k\, U(L_0)_{m_i-k}\, L_{-\alpha}^k\right).$$

Applying the projection ψ gives

$$\psi(f^{m_i+1} e^{m_i+1}) \equiv a_{m_i+1,0}\, [fe]^{m_i+1}$$

$$(\text{modulo } U(L_0)_{m_i}),$$

with $a_{m_i+1,0} > 0$. Thus $[fe]^{m_i+1}$ is congruent, modulo $J(\lambda)$, to an element of $U(L_0)_{m_i}$.

Because the $\{[fe]\}$ run over a basis for L_0, the PBW theorem for $U(L_0)$ and an argument like that on p. 189 of [15] show that the ordered monomials in this basis, of degree at most $m+1$ in each basis element, linearly generate $U(L_0)$ modulo $J(\lambda)$. This gives the bound on the dimension of $U(L_0)/J(\lambda)$.

All the other conclusions are consequences of this and of Theorem 1. Thus Proposition 10 is proved.

If we adjoin $J(\lambda)$ the elements $h_i - \lambda(h_i)1$ in $U(L_0)$ to generate a larger ideal $K(\lambda)$, then it is clear that each irreducible module for $U(L_0)/K(\lambda)$ is finite-dimensional and λ-admissible. Conversely, $K(\lambda)$ annihilates each such module. A refinement of our bound on dimensions in Proposition 10 gives

$$\dim(U(L_0)/K(\lambda)) \le (m+1)^{[L_0:F] - [T:F]}.$$

If M is any (right) module for $U(L_0)/K(\lambda)$, consider M as $U(L_0)$-module, hence as L_0-module: Each $h \in T$ acts as the scalar $\lambda(h)$, and $[L_0 L_0]$ is either zero or semisimple, with T central in L_0. Thus if $L_0 = T + [L_0 L_0]$, M is necessarily a completely reducible L_0-module, therefore a semisimple module for $U(L_0)/K(\lambda)$. That is, _if_ _$L_0 = T + [L_0 L_0]$, then $U(L_0)/K(\lambda)$ is semisimple._

Beginning with Chapter V, we shall study in detail $U(L_0)$-modules annihilated by the ideal $K(\lambda)$ for all reduced central simple Lie algebras and for certain "fundamental" λ.

CHAPTER II

BEHAVIOR UPON SPLITTING. CARTAN MULTIPLICATION

Although the methods of construction of our modules will be rational over F, we shall now and then use a few bits of information from the split case in order to show that the constructions are exhaustive. Those of a general nature are summarized here, particularly for the purpose of showing the exhaustiveness of a relativized "Cartan multiplication".

1. MODULES AND FIELD EXTENSION.

Let L be central simple over F, as in Chapter I, T a maximal F-split torus in L, H a Cartan subalgebra of L containing T (to see that such an H exists, one need only take H to be a commutative subalgebra containing T and such that for all ad h, $h \in H$, are semisimple, H being maximal with these properties; cf. Chapter I of [26]). Let L_0 ($\supseteq H$) be the centralizer of T in L.

Let K be a finite extension field of F (which may be assumed galois over F, as needed) that contains all characteristic roots of all ad h_i, h_i in a basis for H. Then H_K is a splitting Cartan subalgebra for L_K, in the sense of [15], Chapter IV. Let Σ be the set of roots of L_K relative to H_K. Then $(L_0)_K = H_K + \Sigma_{\gamma \in \Sigma_0}(L_K)$, where

$$\Sigma_0 = \{\gamma \in \Sigma \mid \gamma(T) = 0\}.$$

We assume an ordering has been established on the roots of L relative to T, and introduce an ordering on the roots of L_K relative to H_K consistent with the earlier one; that is, for $\gamma_1, \gamma_2 \in \Sigma$, we require that if $\gamma_1|T > \gamma_2|T$ in the ordering of roots (and 0) relative to T, then $\gamma_1 > \gamma_2$ in our ordering of Σ. That this compatibility may always be achieved is shown e.g., in [22], §2, where it is also shown that we may assume a fundamental system $\gamma_1, \ldots, \gamma_\ell$ of (positive) roots relative to H_K, whose restrictions to T contain a fundamental system $\alpha_1, \ldots, \alpha_r$ for L relative to T. Let $\pi_1, \ldots, \pi_\ell$ be the fundamental dominant weights associated with $\gamma_1, \ldots, \gamma_\ell$, i.e., $2(\pi_i, \gamma_j) = \delta_{ij}(\gamma_j, \gamma_j)$ for all i,j, where the form is that induced on H_K^* by the restriction to H_K of the Killing form of L_K.

If P denotes the set of positive roots of L relative to T, we have $N_K = \sum_{\alpha \in P} (L_\alpha)_K = \sum_{\substack{\gamma \in \Sigma \\ \gamma|T \in P}} (L_K)_\gamma$, where N is as in Chapter I, and likewise for N^-.

Let V be an irreducible (hence absolutely irreducible, by [15]) right L_K-module, and let $V^+ = \{u \in V \mid u\, N_K = 0\} = \{u \in V | uN = 0\}$. Since $(L_0)_K$ normalizes N_K, V^+ is an $(L_0)_K$-submodule.

<u>PROPOSITION II.1.</u> V^+ <u>is an irreducible</u> $(L_0)_K$<u>-module, if</u> V <u>is an irreducible</u> L_K<u>-module</u>.

<u>PROOF</u>. From [15] we know that V has a highest weight vector v_0 and is a sum of weight-spaces. By the compatibility of orderings, $v_0 \in V^+$. Because $L \subseteq L_K$ and [K:F] is finite, V is a finite-dimensional L-module. Thus we may consider the L_0-submodule $v_0\, U(L_0)$ of V generated by v_0. Since L_0 normalizes N, we have $v_0\, U(L_0) \subseteq V^+$, and likewise the K-span $Kv_0\, U(L_0) \subseteq V^+$. We next show

(1) $$V^+ = K\, v_0\, U(L_0)\ .$$

From the decomposition (6) of $U(L)$ used in defining the map ψ of §I.2, we have

$$V = v_0\, U(L_K) = K\, v_0\, U(L) = K\, v_0\, U(L_0) + K\, v_0\, U(L_0 + N^-)\, N^-\ .$$

Now N^- has a basis of elements belonging to roots $-\alpha$, $\alpha \in P$, from which one sees that $K\, v_0\, U(L_0 + N^-)\, N^-$ consists of combinations of weight vectors for T belonging to weights of the form $\lambda - \alpha_{i_1} \dots \alpha_{i_t}$, $t > 0$, $\alpha_{i_j} \in P$, where $\lambda = \pi|T$, π the highest weight of V. Meanwhile all elements of $K\, v_0\, U(L_0)$ belong to the weight λ. It follows that the sum

(2) $$V = K\, v_0\, U(L_0) + K\, v_0\, U(L_0 + N^-)\, N^-$$

is direct. Thus (1) will follow if we show that $W = V^+ \cap (K\, v_0\, U(L_0 + N^-)N^-)$ is zero. But clearly W is an $(L_0)_K$-submodule of V^+, from which it follows as in Proposition I.1 that $KW\, U(N^-)$ is an L_K-submodule of V. As T-module, $KW\, U(N^-)$ is the sum of weight-spaces for weights $\lambda - \alpha_{i_1} \dots \alpha_{i_t}$, $t > 0$, $\alpha_{i_j} \in P$; in particular, λ is <u>not</u> a weight. But the irreducibility of the L_K-module V implies that $K\, W\, U(N^-)$ is either V or 0, and λ is a weight of V. Hence $W = 0$, and (1) is proved.

Now let X be a non-zero $(L_0)_K$-submodule of V^+. As with W above, $XU(N^-) = V$ by irreducibility of V as L_K-module. Thus λ is a (T-) weight of $X\, U(N^-)$, so must be a weight of X. Accordingly we may take $v_0 \neq 0$ in X, v_0 belonging to the weight λ for T. It will be noted that (1) above followed from the conditions: a) $0 \neq v_0 \in V^+$; b) v_0 belongs to the T-weight λ. Thus X contains $Kv_0\, U(L_0) = V^+$, by (1), and

$X = V^+$. The proposition is proved.

PROPOSITION II.2. i) Any irreducible L_K-module is an isotypic L-module (i.e., all irreducible L-submodules are isomorphic).

ii) For every irreducible L-module U there is an irreducible L_K-module V such that U is an irreducible L-submodule of V.

iii) Let $\{U_i \mid i \in I\}$ be a family of irreducible L-modules, and let U be an irreducible L-module such that every irreducible L_K-submodule of U_K occurs as an L_K-submodule of some $(U_i)_K$. Then U is L-isomorphic to one of the U_i.

PROOF. All the conclusions rest ultimately on the complete reducibility of modules for L and for L_K and on the uniqueness of their irreducible summands. For i), note that if V is an irreducible L_K-module, and if $V_1, \ldots, V_n$ are irreducible L-submodules with $V = \oplus_{i=1}^n V_i$, then for each $\xi \in K$ the map $v \to \xi v$ of V into V is an L-endomorphism, so maps each V_i into an irreducible L-submodule isomorphic to V_i (or into 0, if $\xi = 0$). Thus the sum of those V_i which are L-isomorphic to V_1 (say) is an L_K-submodule of V, so is V. This proves i). Recalling Proposition I.1, we know that $V^+ \cap V_i$ is the highest (T-) weight-space of the irreducible L-module V_i, so that i) implies that $V^+ = \oplus_{i=1}^n (V^+ \cap V_i)$ is a sum of n isomorphic irreducible L_0-submodules.

To see ii), consider the L_K-module $U_K = K \otimes_F U$. As L-module, U_K is the direct sum of [K:F] copies of U, so every irreducible L-submodule is isomorphic to U. Thus if V is an irreducible L_K-submodule of U_K, every irreducible L-submodule of V is isomorphic to U, and V satisfies ii).

The assertion iii) follows from the reasoning of ii). Let V be an irreducible L_K-submodule of U_K. The hypotheses imply that V is isomorphic to an irreducible L_K-submodule V_i of some $(U_i)_K$. By ii), every irreducible L-submodule of V is isomorphic to U and every irreducible L-submodule of V_i is isomorphic to U_i. Thus U and U_i are L-isomorphic, and iii) is proved.

2. CARTAN MULTIPLICATION.

Let V be an irreducible L_K-module, with highest weight π and highest weight vector v_0, in the setting of §1. Let V', π', v_0' be a second such. Let V^+, V'^+ be the respective subspaces annihilated by N, and let $V \cong_L \oplus_{i=1}^n V_i$, $V' \cong_L \oplus_{j=1}^m V_j'$, where the V_i and V_j' are irreducible L-modules. The Cartan product of V and V' is the L_K-submodule of $V \otimes_K V'$ generated by $v_0 \otimes v_0'$, and is necessarily irreducible by Proposition I.1 in the split case. (Every H_K-weight of $V \otimes_K V'$ is of the form $(\pi + \pi') - \Sigma_{i=1}^{t} \beta_i$, where $\beta_i \in \Sigma$ are positive roots, and the space in $V \otimes_K V'$ of weight $\pi + \pi'$ is $K(v_0 \otimes v_0')$.)

Evidently $W = K(v_0 \otimes v_0')\, U(L_0)$ is an $(L_0)_K$-submodule of $(V \otimes_K V')^+$, the subspace of $V \otimes_K V'$ annihilated by N. With $\lambda = \pi|T$, $\lambda' = \pi'|T$, each $t \in T$ acts on W as scalar multiplication by $(\lambda + \lambda')(t)$. From (1) and (2) of Proposition 1 and their counterparts for V', one sees that $V^+ \otimes_K V'^+$ is the subspace of $V \otimes_K V'$ of T-weight $\lambda + \lambda'$.

The center C of L_0 is contained in H, and the elements of H act semisimply in representations of L_K, hence in representations of L. As reductive subalgebra of L, $L_0 = C \oplus [L_0 L_0]$, where $[L_0 L_0]$ is semisimple. The system of roots for $[L_0 L_0]_K$ relative to its splitting Cartan subalgebra $H_K \cap [L_0 L_0]_K$ is the set of restrictions of Σ_0, in the notation of §1. (cf. [22]).

By the proof of Proposition 1, W is an irreducible $(L_0)_K$-module. It is contained in

$$V^+ \otimes_K V'^+ = (\sum_i (V^+ \cap V_i)) \otimes_K (\sum_j (V'^+ \cap V_j')).$$

(One cannot "distribute" the tensor product here, because the summands are not K-subspaces, in general.) Since the elements of C act semisimply in $V \otimes_K V'$, W is a completely reducible L_0-module, and the proof of i) of Proposition 2 shows that W is an <u>isotypic</u> L_0-module. If W_0 is an irreducible L_0-submodule of W, then W_0 belongs to the weight $\lambda + \lambda'$ for T. From Propositions 1 and 2 of Chapter I, the L-submodule of $V \otimes_K V'$ generated by W_0 is $W_0\, U(N^-)$, is finite-dimensional and irreducible, and its isomorphism-class is independent of the choice of W_0 within W. We refer to $W_0\, U(N^-)$ as <u>the irreducible L-module obtained from V and V' by Cartan multiplication</u>. (Here, of course, T, H, K and the ordering of the roots are fixed.)

If M and M' are two irreducible L-modules, with respective highest T-weights μ, μ', so that $M^+ = M_\mu$, $M'^+ = M'_{\mu'}$, are the subspaces annihilated by N, we say that an irreducible L-module M'' <u>is obtained from M and M'</u> by Cartan multiplication if M'' is isomorphic to an L-submodule of $M \otimes_F M'$ generated by an irreducible L_0-submodule of $M^+ \otimes_F M'^+$ (which is a completely reducible L_0-module, as above). Since $M^+ \otimes_F M'^+$ need not be L_0-isotypic, M'' is not usually unique.

In [4], Cartan observed that, once one has an irreducible L_K-module for each of the fundamental dominant weights $\pi_1, \ldots, \pi_\ell$ (<u>i.e.</u>, a module V_i having π_i as highest weight), then the irreducible L_K-module of highest weight $\pi = \sum_i m_i \pi_i$ $(m_i \in Z^+)$ is the repeated Cartan product of V_i's, the factor V_i entering m_i times; in other words, one forms the tensor product over K of m_1 copies of V_1, m_2 of V_2, etc., and takes the submodule generated by the decomposable tensor whose factors are the highest weight vectors. The next proposition gives the analogue of Cartan's observation in our relative setting.

PROPOSITION II.3. *With the notations and conventions of this chapter, let λ be a dominant integral function on T. Then all the irreducible L-modules with highest weight λ are obtained by repeated Cartan multiplication from those with highest weight $\pi_i|T$, $1 \le i \le \ell$. More precisely, one considers all expressions $\Sigma_{i=1}^{\ell} m_i \pi_i$ $(m_i \in Z^+)$ whose restrictions to T are λ; for each such expression one forms all tensor products of irreducible L-modules involving m_i factors of highest weight $\pi_i|T$ for each i (not necessarily all isomorphic for the same i, and not necessarily non-isomorphic for different i's, if $\pi_i|T = \pi_j|T$); from each such tensor product one extracts the submodules generated by irreducible L_0-submodules of the tensor product of the highest weight spaces. The results are exactly (with repetitions) the irreducible L-modules of highest weight λ.*

PROOF. Let M be an irreducible L-module of highest weight λ. Then M is an irreducible L-summand of any irreducible L_K-summand of $K \otimes_F M$, and $K \otimes_F M^+ = (K \otimes_F M)^+ = \Sigma_i V^{(i)+}$, where the $V^{(i)}$ are the irreducible L_K-summands of $K \otimes_F M$. It follows that the highest weight of each of the $V^{(i)}$ has a restriction to T equal to λ. Thus each of these highest weights has an expression $\Sigma_{j=1}^{\ell} m_j^{(i)} \pi_j$, $m_j^{(i)} \in Z^+$, $(\Sigma_{j=1}^{\ell} m_j^{(i)} \pi_j)|T = \lambda$, as in the Proposition.

Let $V_1, \ldots, V_\ell$ be the irreducible L_K-modules with highest weights $\pi_1, \ldots, \pi_\ell$, respectively, and let $V_k \cong_L \Sigma V_{kj}$ be their decompositions into irreducible L-modules V_{kj}, all those for a given k having highest weight $\pi_k|T$. Now pick $j_1, \ldots, j_\ell$ arbitrarily with $V_{1,j_1}, \ldots, V_{\ell,j_\ell}$ defined as above, and form the tensor product W over F of $m_1^{(i)}$ copies of V_{1,j_1}, $m_2^{(i)}$ of V_{2,j_2}, $\ldots$, $m_\ell^{(i)}$ of V_{ℓ,j_ℓ}, for some fixed i. To show the main assertion of Proposition 3 we show that the subspace of W belonging to the (highest) weight λ has an L_0-summand isomorphic to M^+.

We have seen that $\otimes_K(\otimes_K^{m_k^{(i)}} V_k^+) = \otimes_K(\otimes_K^{m_k^{(i)}} (\Sigma_j(V_k^+ \cap V_{kj})))$ is the subspace of $\otimes_K(\otimes_K^{m_k^{(i)}} V_k)$ belonging to the T-weight λ. Now the F-linear mapping from $\otimes_F(\otimes_F^{m_k^{(i)}} V_k^+)$ to $\otimes_K(\otimes_K^{m_k^{(i)}} V_k^+)$ sending $v_1 \otimes_F \cdots \otimes_F v_s$ to $v_1 \otimes_K \cdots \otimes_K v_s$ is evidently an L_0-homomorphism of the former space onto the latter. By the split theory of Cartan multiplication, the L_K-submodule of $\otimes_K(\otimes_K^{m_k^{(i)}} V_k)$ generated by the tensor product of the highest weight vectors (for H_K) is the irreducible L_K-module of highest weight $\sum_{k=1}^{\ell} m_k^{(i)} \pi_k$, and is therefore L_K-isomorphic to $V^{(i)}$, which has M as L-submodule. Thus M is an L-submodule of $\otimes_K(\otimes_K^{m_k^{(i)}} V_k)$, and M^+ is an L_0-submodule of $\otimes_K(\otimes_K^{m_k^{(i)}} V_k^+)$.

The fact that $\otimes_K(\otimes_K^{m_k^{(i)}} V_k^+)$ is a homomorphic image of the completely reducible L_0-module $\otimes_F(\otimes_F^{m_k^{(i)}} V_k^+)$ implies that M^+ is an irreducible L_0-submodule of

$\oplus (\otimes_F^{\ell}{}_{k=1} ((V_k^+ \cap V_{k,t_1}) \otimes_F (V_k^+ \cap V_{k,t_2}) \otimes_F \cdots \otimes_F (V_k^+ \cap V_{k,t_m}))$, where $m = m_k^{(i)}$ for given k, and where $(t_1, \ldots, t_m)$ runs over all sequences of length m from $\{1, \ldots, r_k\}$ where r_k is the number of irreducible L-summands V_{kj} of V_k (all of them L-isomorphic). That is, M^+ is an irreducible L_0-submodule of

(3) $$\oplus (\otimes_F^{\ell}{}_{k=1} (V_{k,t_1}^+ \otimes_F \cdots \otimes_F V_{k,t_m}^+))$$

with the same conventions.

Each term in the direct sum is the part of the corresponding tensor product of $V_{k,t}$'s over F belonging to the weight λ , and all the terms in this sum,

$$\oplus(\otimes_F^{\ell}{}_{k=1} (V_{k,t_1} \otimes_F \cdots \otimes_F V_{k,t_m})) ,$$

are isomorphic L-modules. Thus all terms of (3) are isomorphic L_0-modules, and each must contain a submodule isomorphic to M^+. This L_0-submodule in turn generates an L-module isomorphic to M. In particular this is the case when all those t_ν associated with k are taken equal to j_k, as proposed early in the proof.

The remainder of Proposition 3 says simply that submodules generated by irreducible L_0-submodules of tensor products of highest weight-spaces of irreducible L-modules whose highest weights add up to λ are irreducible L-modules of highest weight λ . This follows at once from Proposition I.1.

A caution as to what Proposition 3 does <u>not</u> claim: It does not assert or imply that if $\lambda = (\Sigma\, m_i\, \pi_i)|T = (\Sigma\, n_i\, \pi_i)|T$ we can ignore one of these expressions and hope to construct all modules of highest weight λ by our process, applied to the other expression. (Note that the weight $\Sigma m_j^{(i)} \pi_j$ of our proof arose from the module M, not from its highest weight.) For example, it will be seen in the Appendices that if $\mathcal{D}$ is a central (associative) division algebra of degree $d \geq 2$, and if $L = s\ell(n, \mathcal{D})$, $n > 3$, then a fundamental system of roots for L_K is $\gamma_1, \ldots, \gamma_{nd-1}$, of type A:

Denoting by γ' the restriction of γ to T, we have $\gamma_j' = 0$ if $d \nmid j$, while a fundamental system relative to T is $\gamma_d', \ldots, \gamma_{(n-1)d}'$, also of type A:

Among the fundamental weights are

$$\pi_1 = \frac{1}{nd}\ ((nd-1)\gamma_1 + (nd-2)\gamma_2 + \ldots + \gamma_{nd-1}),$$
$$\pi_2 = \frac{1}{nd}\ ((nd-2)\gamma_1 + 2(nd-2)\gamma_2 + 2(nd-3)\gamma_3 + \ldots + 2\gamma_{nd-1}),$$

so that $\pi_2|T = 2\pi_1|T$. Choice of the expression $\lambda = \pi_2|T$ would have resulted in only one irreducible L-module, that occurring in the irreducible L_K-module of highest weight π_2. On the other hand, choice of the expression $(\pi_1 + \pi_1)|T$ will be seen to yield two irreducible modules by Cartan multiplication of the unique irreducible L-module with highest weight $\pi_1|T$. (See Chapter V.)

CHAPTER III

MAPPINGS SATISFYING SYMMETRIC IDENTITIES

In Chapters V and VI, the irreducible λ-admissible L_0-modules, for L of relative types A and C, will be determined for a set of values of λ that is adequate in the sense of Cartan multiplication. There the essential condition for λ-admissibility will be reduced to the study of certain combinations derived from a mapping of an associative division algebra into the F-endomorphisms of our module. This chapter concentrates on the combinatorial formalism involved and on the realization, by a symmetric power, of an algebra that in many of these cases identifies with $U(L_0)/K(\lambda)$, of §1.4. In particular, an identity for the canonical Lie morphism of an associative algebra into (the symmetric elements of) its k-fold tensor power is central to our studies. This identity yields a number of consequences in identities for classical algebra, which we shall indicate.

1. SOLUTION OF THE BASIC RECURSION.

LEMMA III.1. *Let A be an associative algebra with unit; let S be a subset of A such that if a_1 and a_2 are commuting elements of S, then $a_1a_2 \in S$, and let ρ be a mapping of S into an associative algebra B with unit such that $\rho(a_1)$ and $\rho(a_2)$ commute whenever a_1 and a_2 commute. For each Cartesian power S^t of S let a mapping f_t of S^t to B be given, such that*

(i) $f_1(a) = \rho(a)$ *for all* $a \in S$;

(ii) *If* $a_1,\ldots,a_{t+1}$ *is a sequence of commuting elements of* S,

$$f_{t+1}(a_1,\ldots, a_{t+1}) = \sum_{i=1}^{t+1} \rho(a_i)f_t(a_1,\ldots,\hat{a}_i,\ldots,a_{t+1})$$

(1)
$$-2 \sum_{i<j} f_t(a_ia_j,a_1,\ldots, \hat{a}_i,\ldots, \hat{a}_j,\ldots, a_{t+1})$$

for every $t \geq 1$, *where the circumflex indicates an omitted argument.*

Then for every t *and every commuting sequence* $a_1,\ldots, a_t \in S$,

$$f_t(a_1,\ldots, a_t) =$$

$$(2) \qquad t! \sum^{(1)}_{\nu_1+\ldots+\nu_m=t} \prod_{i=1}^{m} (-1)^{\nu_i-1}(\nu_i-1)! \; \Sigma^{(2)} \rho(a_{i_1}\cdots a_{i_{\nu_1}}) \cdots$$

$$\cdots \rho(a_{i_{\nu_1+\ldots+\nu_{m-1}+1}} \cdots a_{i_{\nu_1+\ldots+\nu_m}}) .$$

The sum $\Sigma^{(1)}$ *is over all partitions* $1 \le \nu_1 \le \ldots \le \nu_m$ *of* t; *for each such partition,* $\Sigma^{(2)}$ *is the sum over all permutations* $(i_1,\ldots, i_t)$ *of* $(1,\ldots, t)$, *subject to the following restrictions:*

(α) For each $j < m$, $i_{\nu_1+\ldots+\nu_j+1} < i_{\nu_1+\ldots+\nu_j+2} < \cdots$ $i_{\nu_1+\ldots+\nu_{j+1}}$ (with $\nu_0 = 0$) --- the a_i in each block are in order;

(β) *If* $\nu_k = \nu_\ell$ *and* $k < \ell$, *then* $i_{\nu_1+\ldots+\nu_{k-1}+1} < i_{\nu_1+\ldots+\nu_{\ell-1}+1}$; *that is, given two blocks of equal size, the one appearing first contains an* a_i *that precedes all those in the later block.*

PROOF. For $t = 1$, the formula (2) reduces to $f_1(a) = \rho(a)$. To carry out the inductive step, let $a_1,\ldots, a_{t+1}$ be commuting elements of S; substituting the form of (2) assumed to hold for t in (1), we obtain

$$f_{t+1}(a_1,\ldots, a_{t+1}) =$$

$$(3) \qquad \sum_{i=1}^{t+1} t! \sum_{\nu_1+\ldots+\nu_m=t} \prod_{j=1}^{m} (-1)^{\nu_j-1} (\nu_j-1)! \; \Sigma^{(2)} \rho(a_i)\rho(a^{(i)}_{i_1}\cdots a^{(i)}_{i_{\nu_1}})$$

$$\cdots \rho(a^{(i)}_{i_{\nu_1+\ldots+\nu_{m-1}+1}} \cdots a^{(i)}_{i_{\nu_1+\ldots+\nu_m}})$$

$$-2 \sum_{i<j} t! \; \Sigma^{(1)} \prod_{k=1}^{m} (-1)^{\nu_k-1} (\nu_k-1)! \; \Sigma^{(2)} \rho(a^{(i,j)}_{i_1} \cdots a^{(i,j)}_{i_{\nu_1}})$$

$$\cdots \rho(a^{(i,j)}_{i_{\nu_1+\ldots+\nu_{m-1}+1}} \cdots a^{(i,j)}_{i_{\nu_1+\ldots+\nu_m}}) .$$

Here $a^{(i)}_k = a_k$ if $k < i$ and $a^{(i)}_k = a_{k+1}$ if $k \ge i$; $a^{(i,j)}_1 = a_i a_j$; $a^{(i,j)}_k = a_{k-1}$ if $1 < k \le i$; $a^{(i,j)}_k = a_k$ if $i < k < j$; $a^{(i,j)}_k = a_{k+1}$ if $j \le k \le t$.

Every term in the expansion of (3) is a product

$$(4) \qquad \rho(a_{j_1} \cdots a_{j_{\mu_1}})\rho(a_{j_{\mu_1+1}} \cdots a_{j_{\mu_1+\mu_2}}) \cdots \rho(a_{j_{\mu_1+\ldots+\mu_{r-1}+1}} \cdots a_{j_{\mu_1+\ldots+\mu_r}})$$

for some partition $t + 1 = \mu_1 + \ldots + \mu_r$ and for some permutation

$(j_1,\dots, j_{t+1})$ of $(1,2,\dots, t+1)$, with an appropriate coefficient. By the commutativity of the factors, we may arrange those a_k to whose product a single ρ applies in increasing order of indices and may arrange the r factors which are images under ρ so that $\mu_1 \le \mu_2 \le \dots \le \mu_r$, and furthermore so that if $\mu_s = \mu_{s+1}$ then the first index in the s-th block precedes the first index in the (s+1)-th block. We collect all the terms of (3) which result in (4) upon such rearrangements, where (4) is assumed in the canonical rearranged form, and determine the resulting coefficient.

When all $\mu_k = 1$, the term (4) is $\rho(a_1)\dots\rho(a_{t+1})$, and never occurs in the second summation of (3). For each of the $t+1$ possible choices of i our term occurs in the first summation of (3) exactly when $\nu_1 = \dots = \nu_t = 1$, in which case the corresponding summation $\Sigma^{(2)}$ is $\rho(a_i)\rho(a_1)\dots\widehat{\rho(a_i)}\dots\rho(a_{t+1}) = \rho(a_1)\dots\rho(a_i)\dots\rho(a_{t+1})$. The factor $\Pi\,(-1)^{\nu_j-1}(\nu_j-1)!$ is equal to 1, and the coefficient of our term is $\sum_{i=1}^{t+1} t! = (t+1)!$. This agrees with the coefficient of $\rho(a_1)\dots\rho(a_{t+1})$ in (2) when t is replaced by $t+1$.

Now assume some $\mu_s > 1$ in (4). The first sum of (3) will contribute to the coefficient of (4) only if $\mu_1 = 1$; in this case $\mu_2 \le \mu_3 \le\dots\le \mu_r$ is a partition of t, and the contribution of the first summation of (3) to the coefficient of (4) comes from terms

$$t!\ \prod_{j=2}^{r}(-1)^{\mu_j-1}(\mu_j-1)!\ \rho(a_i)\ \rho(a_{i_1}^{(i)}\dots a_{i_{\mu_2}}^{(i)})\dots .$$

If exactly n of the μ_j's are equal to 1, the product (4) will appear whenever i takes on one of the corresponding values $j_1, j_2,\dots, j_n$, making a total contribution, from the first sum of (3), of

$$n\cdot t!\ \prod_{j=2}^{r}(-1)^{\mu_j-1}(\mu_j-1)! = n\cdot t!\ \prod_{j=1}^{r}(-1)^{\mu_j-1}(\mu_j-1)!$$

to the coefficient of (4) in $f_{t+1}(a_1,\dots, a_{t+1})$.

Next consider the second sum of (3) which makes a contribution to the coefficient of (4) exactly when the selected pair $i < j$ of indices is contained in one block in the partition, associated with (4), of the index-set $\{1,\dots, t+1\}$. Fixing such a pair $i < j$, let it occur in the $\gamma(i,j)$-th block, of size $\mu_{\gamma(i,j)} \ge 2$. Then (4) arises from a partition of t into r terms where all blocks but this one have the same size as in the partition $t+1 = \mu_1 + \dots + \mu_r$, while one block, corresponding to the factor in (4) containing $a_1^{(i,j)} = a_i a_j$, has size $\mu_{\gamma(i,j)}-1$. The corresponding coefficient in (3) is

$$-2t!\left(\prod_{\substack{s=1\\ s\neq\gamma(i,j)}}^{r}(-1)^{\mu_s-1}(\mu_s-1)!\right)\cdot(-1)^{\mu_{\gamma(i,j)}-2}(\mu_{\gamma(i,j)}-2)!$$

$$= 2t!\,(\mu_{\gamma(i,j)}-2)!\prod_{s\neq\gamma(i,j)}(\mu_s-1)!\prod_{s=1}^{r}(-1)^{\mu_s-1}.$$

For fixed γ there is such a contribution whenever both indices i and j are in the γ-th block, a total of

$$\binom{\mu_\gamma}{2}2t!(\mu_\gamma-2)!\prod_{s\neq\gamma}(\mu_s-1)!\prod_{s=1}^{r}(-1)^{\mu_s-1}$$

$$= t!\,\mu_\gamma\prod_{s=1}^{r}(-1)^{\mu_s-1}(\mu_s-1)!$$

Finally, note that the second member of (4) makes such a contribution to the coefficient of (4) for every block of size $\mu_\gamma > 1$ in the partition $\mu_1 \leq \ldots \leq \mu_r$ of $t + 1$, a total contribution of

$$t!\prod_{s=1}^{r}(-1)^{\mu_s-1}(\mu_s-1)!\sum_{\mu_\gamma>1}\mu_\gamma = (t+1-n)t!\prod_{s=1}^{r}(-1)^{\mu_s-1}(\mu_s-1)!,$$

where n is the number of $\mu_s = 1$. Combining this with the contribution from the first sum of (3), we obtain a coefficient of

$$(t+1)!\prod_{s=1}^{r}(-1)^{\mu_s-1}(\mu_s-1)!$$

for the term (4) in $f_t(a_1,\ldots, a_{t+1})$. The induction is complete.

<u>COROLLARY III.2.</u> <u>In the notations of the Lemma,</u>

$$f_t(a,a,\ldots, a) =$$

$$(5)\qquad t!\sum_{P\leftrightarrow(s_1,\ldots,s_t)}\prod_{i=1}^{t}((-1)^{i-1}(i-1)!)^{s_i}\cdot\frac{t!}{1!^{s_1}2!^{s_2}\ldots t!^{s_t}}\cdot$$

$$\frac{1}{s_1!s_2!\ldots s_t!}\,\rho(a)^{s_1}\rho(a^2)^{s_2}\ldots\rho(a^t)^{s_t}.$$

<u>Here</u> $P \leftrightarrow (s_1,\ldots, s_t)$ <u>means that</u> s_1 <u>of the blocks in the partition</u> P <u>of</u> t <u>have size</u> 1, s_2 <u>have size</u> $2,\ldots,$ s_t <u>have size</u> t; <u>the sum is taken over all partitions</u> P <u>of</u> t.

The Corollary is obtained by noting that if we collect the terms in (2), for $f_t(a,a,\ldots, a)$, corresponding to the partition P: $1 \leq \nu_1 \leq \nu_2 \leq \ldots \leq \nu_m$ of t, we find that the term involving $\rho(a^{\nu_1})\rho(a^{\nu_2})\ldots\rho(a^{\nu_m})$ occurs

$$\binom{t}{\nu_1,\ \nu_2,\ldots,\ \nu_m}\frac{1}{s_1!\ldots s_t!}$$

times.

The factor $\prod_{i=1}^{t}(-1)^{s_i(i-1)}$ will be recognized as the sign $\operatorname{sgn}(C(P))$ of any permutation in the conjugacy class of the symmetric group S_t associated with P (i.e., whose cycle-structure is determined by P). The factor

$$\prod_{i=1}^{t}(i-1)!^{s_i}\ \frac{t!}{1!^{s_1}\cdots t!^{s_t}}\ \frac{1}{s_1!\cdots s_t!} = \frac{t!}{1^{s_1}2^{s_2}\cdots t^{s_t}s_1!\cdots s_t!}$$

is the cardinality $|C(P)|$ of this conjugacy class. Thus (5) may be more concisely written as

$$f_t(a,a,\ldots,a) = t!\sum_{P\leftrightarrow(s_1,\ldots,s_t)}\operatorname{sgn}(C(P))\,|C(P)|\,\rho(a)^{s_1}\rho(a^2)^{s_2}\cdots\rho(a^t)^{s_t}. \tag{6}$$

2. THE CANONICAL EXAMPLE.

Let A be an associative algebra with unit over F, and let the algebra B over F be the k-th tensor power $\otimes^k A$. Let $\rho = \rho_k : A \to B$ be the mapping $a \to (a\otimes 1\otimes\ldots\otimes 1) + (1\otimes a\otimes 1\otimes\ldots\otimes 1) +\ldots+ (1\otimes\ldots\otimes 1\otimes a)$ (here $S = A$). Here $\rho_k([ab]) = [\rho_k(a),\rho_k(b)]$, so it is clear that $\rho_k(a)$ and $\rho_k(b)$ commute if a and b do. For $1 \le m \le k$, and for $a_1,\ldots,a_m \in A$, define $g_m(a_1,\ldots,a_m) \in B$ by

$$g_m(a_1,\ldots,a_m) =$$

$$m'\sum_{1\le i_1<\cdots<i_k\le k}\ \sum_{\pi\in S_m} 1\otimes\ldots\otimes 1\otimes \overset{i_1}{a_{\pi(1)}}\otimes\ldots\otimes \overset{i_2}{a_{\pi(2)}}\otimes\ldots\otimes \overset{i_m}{a_{\pi(m)}}\otimes\ldots\otimes 1. \tag{7}$$

For $m > k$, $g_m(a_1,\ldots,a_m)$ is to be $0 \in B$. We claim $\rho = \rho_k$ <u>and</u> $f_t = g_t$ <u>satisfy the recursion</u> (1).

That $g_1 = \rho_k$ and that (1) holds when $t > k$ are clear. If $1 \le t \le k$, then for commuting a_i,

$$\sum_{i=1}^{t+1}\rho_k(a_i)g_t(a_1,\ldots,\hat{a}_i,\ldots,a_{t+1}) - 2\sum_{i<j}g_t(a_ia_j,a_1,\ldots,\hat{a}_i,\ldots,\hat{a}_j,\ldots a_{t+1})$$

$$= \sum_{i=1}^{t+1} t!\sum_{j=1}^{k}\ \sum_{\substack{1\le i_1<\ldots<i_t\le k\\ (\text{all } i_\ell \ne j)}}\ \sum_{\pi\in S\{1,\ldots,\hat{i},\ldots t+1\}} 1\otimes\ldots\otimes \overset{i_1}{a_{\pi(1)}}\otimes\ldots\otimes \overset{j}{a_i}\ldots\otimes \overset{i_t}{a_{\pi(t+1)}}\otimes\ldots\otimes 1$$

$$+\sum_{i=1}^{t+1} t!\sum_{j=1}^{k}\ \sum_{\substack{1\le i_1<\ldots<i_t\le k\\ (j = i_r,\ \text{some } r)}}\ \sum_{\text{one-one maps } \pi:\{1,\ldots,t\}\to\{1,\ldots,\hat{i},\ldots,t+1\}}$$

$$1 \otimes \ldots \otimes a_{\pi(1)}^{i_1} \otimes \ldots \otimes a_i^{i_r} a_{\pi(r)} \otimes \ldots \otimes a_{\pi(t)}^{i_t} \otimes \ldots \otimes 1$$

$$- 2 \sum_{i<j} t! \sum_{1 \le i_1 < \ldots < i_t \le k} \sum_{\pi \in S_t} 1 \otimes \ldots \otimes a_{\pi(1)}^{i_1(i,j)} \otimes \ldots \otimes a_{\pi(t)}^{i_t(i,j)} \otimes \ldots \otimes 1,$$

where in the last sum $a_m^{(i,j)}$ is as in the proof of Lemma 1.

Of the three multiple summations, the second contains each term

$$\text{(8)} \qquad t!\, 1 \otimes \ldots \otimes a_{j_1}^{\mu_1} \otimes \ldots \otimes a_{j_r}^{\mu_r} a_{j_{r+1}} \otimes \ldots \otimes a_{j_{t+1}}^{\mu_t} \otimes \ldots \otimes 1,$$

with $j_r < j_{r+1}$, twice; once for $i = j_r$, $j = \mu_r$, $\{i_1, \ldots, i_t\} = \{\mu_1, \ldots, \mu_t\}$, $\pi(p) = j_p$ if $p < r$, $\pi(p) = j_{p+1}$ if $p \ge r$; and once for $i = j_{r+1}$, $j = \mu_r$, $\{i_1, \ldots, i_t\} = \{\mu_1, \ldots, \mu_t\}$, $\pi(p) = j_p$ if $p \le r$, $\pi(p) = j_{p+1}$ if $p > r$. The third multiple summation contains this term, with coefficient -2, only for

$i = j_r$, $j = j_{r+1}$, $\{i_1, \ldots, i_t\} = \{\mu_1, \ldots, \mu_t\}$, $\pi(r) = 1$,

$\pi(s) = j_s + 1$ if $s < r$ and $j_s < i$; $\pi(s) = j_s$ if $s < r$ and $i < j_s < j$;

$\pi(s) = j_s - 1$ if $s < r$ and $j_s > j$; $\pi(s) = j_{s+1} + 1$ if $s > r$ and $j_{s+1} < i$;

$\pi(s) = j_{s+1}$ if $s > r$ and $i < j_{s+1} < j$; $\pi(s) = j_{s+1} - 1$ if $s > r$ and $j_{s+1} > j$.

No terms (8) occur in the first multiple summation, while all terms of the second and third have the form (8). It follows that the second and third cancel, leaving only the first. But the first is evidently $g_{t+1}(a_1, \ldots, a_{t+1})$ if $t + 1 \le k$ and is vacuous if $t = k$. Thus (1) holds, and Lemma 1 applies to show that (2) holds for commuting $a_1, \ldots, a_t$ when $f_t = g_t$ and $\rho = \rho_k$. For $t = k+1$, we have

$$\begin{aligned} 0 &= g_{k+1}(a_1, \ldots, a_{k+1}) \\ &= (k+1)! \sum\nolimits^{(1)}_{\nu_1 + \ldots + \nu_m = k+1} \prod_{i=1}^{m} (-1)^{\nu_i - 1} (\nu_i - 1)! \sum\nolimits^{(2)} \rho_k(a_{i_1} \ldots a_{i_{\nu_1}}) \\ &\qquad \ldots \rho_k(a_{i_{\nu_1 + \ldots + \nu_{m-1} + 1}} \ldots a_{i_{k+1}}) \end{aligned}$$

whenever $a_1, \ldots, a_{k+1}$ are commuting elements of A, and

$$\begin{aligned} 0 &= g_{k+1}(a, \ldots, a) \\ \text{(9)} \qquad &= (k+1)! \sum_{P \leftrightarrow (s_1, \ldots, s_{k+1})} \operatorname{sgn}(C(P)) |C(P)| \rho_k(a)^{s_1} \rho_k(a^2)^{s_2} \ldots \rho_k(a^{k+1})^{s_{k+1}} \end{aligned}$$

for all $a \in A$.

In general, we shall say that a mapping ρ of A into B as in Lemma 1 *satisfies the* $(k+1)$-th *symmetric identity on* S if for all $a \in S$,

$$\sum_{P \leftrightarrow (s_1,\ldots,s_{k+1})} \operatorname{sgn}(C(P))\,|C(P)|\,\rho(a)^{s_1}\rho(a^2)^{s_2}\cdots\rho(a^{k+1})^{s_{k+1}} = 0.$$

If $S = A$, we say ρ satisfies the (k+1)-th symmetric identity. Thus the mapping $\rho_k: A \to \otimes^k A$ satisfies the (k+1)-th symmetric identity.

3. A CHARACTERIZATION OF $(S^k(A), \rho_k)$.

Evidently the image of ρ_k lies in $S^k(A)$, the subalgebra of $\otimes^k A$ fixed under the canonical action on $\otimes^k A$ of the symmetric group S_k. (In fact, as we shall see below, the image generates $S^k(A)$.) As we have remarked at the beginning of §2, ρ_k is a Lie morphism. Finally, $\rho_k(1) = k\,1$, where the unit elements are those of A resp. $S^k(A)$. The next proposition establishes the universality of the pair $(S^k(A), \rho_k)$ for the properties we have verified.

PROPOSITION III.3. *Let A be a finite-dimensional associative F-algebra with unit. Let k be a fixed positive integer. Then for every associative F-algebra B with unit and for every Lie morphism ρ of A into B such that $\rho(1_A) = k\,1_B$ and such that ρ satisfies the (k+1)-th symmetric identity, there is a unique homomorphism φ of associative algberas with unit such that the diagram*

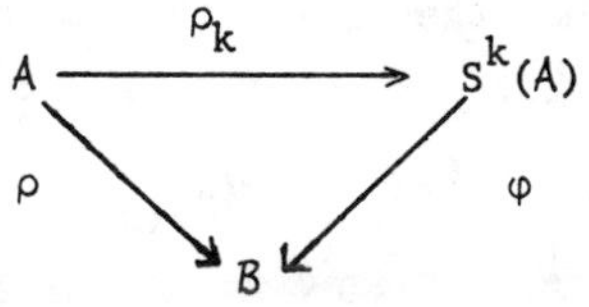

is commutative.

PROOF. Let U be the universal enveloping algebra of the Lie algebra A, $\gamma: A \to U$ the canonical inclusion, J the ideal in U generated by $\gamma(1_A) - k1$ and by the elements

$$\sum_{P \leftrightarrow (s_1,\ldots,s_{k+1})} \operatorname{sgn} C(P)\ |C(P)|\ \gamma(a)^{s_1}\,\gamma(a^2)^{s_2}\ \cdots\ \gamma(a^{k+1})^{s_{k+1}}$$

for all $a \in A$. Then the quotient U/J, with the composite mapping

$$\psi : A \xrightarrow{\ \gamma\ } U \xrightarrow{\ \text{canonical}\ } U/J$$

necessarily has the universal properties claimed for $(S^k(A), \rho_k)$ in the proposition. By §2, we may take $B = S^k(A)$, $\rho = \rho_k$, and obtain a unique homomorphism η of U/J into $S^k(A)$ making commutative the diagram

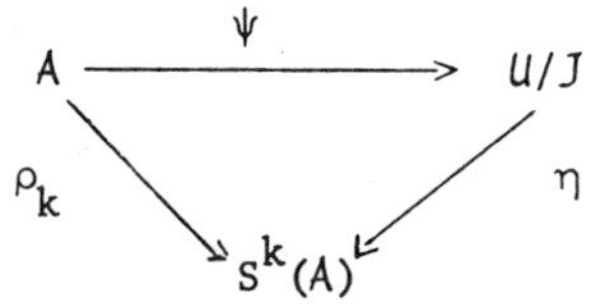

The proof will be completed by showing that η <u>is an isomorphism</u>. This will be achieved by showing:

(i) $\dim (\mathcal{U}/\mathcal{J}) \le \dim S^k(A)$; and

(ii) η maps $\mathcal{U}/\mathcal{J}$ <u>onto</u> $S^k(A)$.

To see (i), let A have dimension n. By the Poincaré-Birkhoff-Witt theorem, the filtration of $\mathcal{U}$ determined by the generating set A has, for each $t \ge 0$, its t-th member $\mathcal{U}_t$ linearly isomorphic to the t-th member $S_t(A)$ in the corresponding filtration of the symmetric algebra $S(A)$ of the vector space A. The dimension of $S_t(A)$ is $\binom{n+t}{t}$, the dimension of the space of polynomials in n indeterminates and of total degree at most t. By taking 1_A to be in a fixed basis for A, we see from $\gamma(1_A)^j \equiv k^j 1_{\mathcal{U}}$ (modulo $\mathcal{J}$) for all $j > 0$ that every element of $\mathcal{U}_t$ is congruent modulo $\mathcal{J}$ to a linear combination of ordered monomials in the remaining $n - 1$ basis elements for A, of degree at most t, thus to an element of a space of dimension $\binom{n-1+t}{t}$.

Now the dimension of $S^k(A)$ is that of $S_k(A)/S_{k-1}(A)$, <u>i.e.</u>, of the <u>homogeneous</u> polynomials of degree k in n variables, or $\binom{n+k-1}{k}$. Therefore (i) can be proved by showing that every element of $\mathcal{U}_{k+1}$ is congruent modulo $\mathcal{J}$ to an element of $\mathcal{U}_k$. When the (k+1)-th symmetric identity is completely polarized, we see that for arbitrary $a_1, \ldots, a_{k+1} \in A$, the term

$$b = \sum_{\pi \in S_{k+1}} \gamma(a_{\pi(1)})\, \gamma(a_{\pi(2)}) \cdots \gamma(a_{\pi(k+1)}) ,$$

corresponding to the partition $(k+1,0,\ldots 0)$ in our generator for $\mathcal{J}$, is congruent modulo $\mathcal{J}$ to an element of $\mathcal{U}_k$. By the P-B-W theorem (or its proof), $b \equiv (k+1)!\, \gamma(a_1) \cdots \gamma(a_{k+1})$ (modulo $\mathcal{U}_k$). Thus $\gamma(a_1) \cdots \gamma(a_{k+1})$ is in $\mathcal{U}_k + \mathcal{J}$ for all $a_1, \ldots, a_{k+1} \in A$, and $\mathcal{U}_{k+1} \subseteq \mathcal{U}_k + \mathcal{J}$. This proves (i).

To show (ii), we note that since $S^k(A)$ is linearly generated by elements

$$\sum_{\pi \in S_k} a_{\pi(1)} \otimes \cdots \otimes a_{\pi(k)}$$

for $a_1, \ldots, a_k \in A$, and since these are combinations of elements $a \otimes \cdots \otimes a$, it will be enough to show each element of this last form is in the subalgebra of $S^k(A)$ generated by $\rho_k(A)$. With $a \in A$ fixed, let

$$b_i = 1 \otimes \cdots \otimes 1 \otimes \overset{i}{a} \otimes 1 \otimes \cdots \otimes 1 \in \otimes^k A,$$

$1 \le i \le k$. By Newton's identities (<u>e.g.</u>, see [14], ex. 4. p. 110), each elementary symmetric function of $b_1, \ldots, b_k$ is a polynomial in

$s_j = \sum_{j=1}^{k} b_i^j$, $j = 1,\ldots, k$. Now $s_j = \rho_k(a^j)$ for each j, and $b_1 \ldots b_k$ is a symmetric function, equal to $a \otimes a \otimes \ldots \otimes a$. The Proposition is proved.

4. APPLICATIONS.

We indicate some applications of the above to identities in associative algebras, especially matrix algebras, and in symmetric functions. Proofs are to be found in [25], which is the main source for this chapter.

By taking A to be $M_n(F)$, the algebra of n by n F-matrices, we have that $S^n(A)$ acts on $\Lambda^n(F^n)$ in such a way that $\rho_n(A)$ acts by multiplication by the trace $t(A)$ of the n by n matrix A. A consequence of the fact that ρ_n satisfies the $(n+1)$-th symmetric identity is then

PROPOSITION III.4. If A is an n by n matrix, then

$$\sum_{P\leftrightarrow (s_1,\ldots,s_{n+1})} \operatorname{sgn} C(P)\ |C(P)|\ t(A)^{s_1} t(A^2)^{s_2} \ldots t(A^{n+1})^{s_{n+1}} = 0. \tag{10}$$

In fact, it is shown in [25] that if $\tau(A) = \frac{1}{n} t(A)$, then $k\tau$ <u>satisfies the</u> $(k+1)$-th <u>symmetric identity if and only if</u> n <u>divides</u> k.

When A is a central simple associative algebra over F, $[A:F] = d^2$, the <u>reduced trace</u> $t(a)$ of $a \in A$ is the element $d\tau(a)$ of F, when $a = \tau(a)\,1 + a_0$, $a_0 \in [AA]$. Upon extension of F to a splitting field, $t(a)$ is the trace of a matrix with entries in the splitting field. By polarization and splitting, Proposition 4 yields

COROLLARY III.5. <u>Let</u> A <u>be a central simple associative algebra over</u> F, $[A:F] = n^2$, <u>and let</u> $t(a) \in F$ <u>be the reduced trace of</u> $a \in A$. <u>Then</u> t <u>satisfies the</u> $(n+1)$-th <u>symmetric identity, and</u> $k\tau$ <u>satisfies the</u> $(k+1)$-th <u>symmetric identity if and only if</u> n <u>divides</u> k. (The last clause is a consequence of the remark following Proposition 4.)

Partial polarization of the $(n+1)$-th symmetric identity in Corollary 5, combined with the nondegeneracy of the symmetric bilinear form $(a,b) = t(ab)$ on A, yields

COROLLARY III.6. <u>Let</u> A <u>be a central simple associative algebra over</u> F, $[A:F] = n^2$, $a \in A$, X <u>an indeterminate over</u> F. <u>Define</u>

$$p_a(X) = \sum_{j=0}^{n} \frac{(-1)^j}{(n-j)!}\ r_{n-j}(a) X^j , \tag{11}$$

<u>where</u> $r_m(a) = \sum_{P\leftrightarrow(s_1,\ldots,s_m)} \operatorname{sgn} C(P)\ |C(P)|\ t(a)^{s_1} t(a^2)^{s_2} \ldots t(a^m)^{s_m}$.

<u>Then</u> $p_a(a) = 0$. <u>Thus</u> $p_a(X) = (-1)^n g_a(X)$, <u>where</u> $g_a(X)$ <u>is the generic minimum polynomial of</u> a (see [1], Chapter 8, 11.)

Specializing Corollary 6 to the case where a is an n by n

diagonal matrix with independent indeterminates $X_1, \ldots, X_n$ on the diagonal, the coefficients of $p_a(X)$ must be (up to sign) the elementary symmetric functions of $X_1, \ldots, X_n$. Thus we have

COROLLARY III.7. Let $X_1, \ldots, X_n$ be indeterminates over F. Let $\sigma_1(X) = \Sigma X_i$, $\sigma_2(X) = \sum_{i<j} X_i X_j, \ldots, \sigma_n(X) = X_1 X_2 \ldots X_n$ be the elementary symmetric functions of $X_1, \ldots, X_n$, and let $t_k(X) = \sum_i X_i^k$. Then for each k, $1 \le k \le n$,

$$k!\,\sigma_k(X) = \sum_{P \leftrightarrow (s_1, \ldots, s_k)} \operatorname{sgn} C(P)\, |C(P)|\, t_1(X)^{s_1} t_2(X)^{s_2} \ldots t_k(X)^{s_k}. \tag{12}$$

In its less concise form, as in §1 of this chapter, the formula (12) appears as an exercise in [21] (p. 458).

CHAPTER IV

STRUCTURE OF SYMMETRIC POWERS

Let A be a central simple associative algebra over the field F, $[A:F] = d^2$. We study the decomposition of the symmetric elements under the permutation-action of the symmetric group S_k in $\otimes^k A$. We also consider central simple involutorial algebras.

1. SUMMARY OF CLASSICAL RESULTS; THE SPLIT CASE.

First suppose $A = \mathrm{End}_F V$, V a d-dimensional vector space over F. Then $\otimes^k A = \mathrm{End}_F(\otimes^k V)$, and $S^k(A)$ is an algebra of F-endomorphisms of $\otimes^k V$. The permutation-action of the finite group S_k in $\otimes^k V$ is completely reducible, and if

$$t = \sum_{i_1,\ldots,i_k} a_{i_1} \otimes a_{i_2} \otimes \ldots \otimes a_{i_k} \in \otimes^k A, \quad a_{i_j} \in A,$$

centralizes the action of S_k, we have for all $\alpha_1,\ldots,\alpha_k$, and for all $v_{\alpha_1},\ldots,v_{\alpha_k} \in V$, $\pi \in S_k$, $(\pi(v_{\alpha_1} \otimes \ldots \otimes v_{\alpha_k}))t = \pi((v_{\alpha_1} \otimes \ldots \otimes v_{\alpha_k})t)$, or

$$(v_{\alpha_{\pi(1)}} \otimes \ldots \otimes v_{\alpha_{\pi(k)}})t = \Sigma\, v_{\alpha_{\pi(1)}} a_{i_{\pi(1)}} \otimes \ldots \otimes v_{\alpha_{\pi(k)}} a_{i_{\pi(k)}} \quad \text{or}$$

$$\Sigma v_{\alpha_{\pi(1)}} a_{i_1} \otimes \ldots \otimes v_{\alpha_{\pi(k)}} a_{i_k} = \Sigma\, v_{\alpha_{\pi(1)}} a_{i_{\pi(1)}} \otimes \ldots \otimes v_{\alpha_{\pi(k)}} a_{i_{\pi(k)}},$$

i.e., t and $\pi(t) = \Sigma\, a_{i_{\pi(1)}} \otimes \ldots \otimes a_{i_{\pi(t)}}$ have the same effect on every $v_{\alpha_{\pi(1)}} \otimes \ldots \otimes v_{\alpha_{\pi(k)}}$. But evidently elements of this form linearly generate $\otimes^k V$, so that $\pi(t) = t$. Conversely it is clear that if $\pi(t) = t$ for all $\pi \in S_k$, then t centralizes the action of S_k on $\otimes^k V$. That is, $S^k(A)$ *is* the centralizer of S_k in $\mathrm{End}_F(\otimes^k V)$.

From the complete reducibility of the action of S_k on $\otimes^k V$ it now follows that $S^k(A)$ is a *completely reducible algebra of* F-*endomorphisms of* $\otimes^k V$, i.e., *a semisimple subalgebra of* $\otimes^k A$, *and that the centralizer of* $S^k(A)$ *is the algebra of* F-*endomorphisms of* $\otimes^k V$ *generated* (linearly) *by* S_k (see [28], p. 95).

The S_k-isotypic components of $\otimes^k V$ (*i.e.*, S_k-submodules maximal with respect to having all their irreducible submodules isomorphic) are clearly stable under $S^k(A)$. If U is one of these components and if

$\mathcal{U} = \mathcal{U}_1 \oplus \dots \oplus \mathcal{U}_r$, where $\mathcal{U}_1, \dots, \mathcal{U}_r$ are isomorphic irreducible S_k-submodules, let $u_1, \dots, u_r$ be fixed S_k-generators for the respective $\mathcal{U}_i$, corresponding under a fixed set of isomorphisms as S_k-modules. Then for each r by r F-matrix (β_{ij}), there is a unique S_k-endomorphism of $\mathcal{U}$ sending each u_i to $\sum_i \beta_{ij} u_j$. It follows that $S^k(A)$ contains the direct sum of full matrix algebras of this kind, one for each type of irreducible S_k-module contained in $\mathcal{V}$.

The classical results of Schur (see, e.g. [28], Chapter 4) show that the group algebra $F[S_k]$ of S_k over F is a direct sum of minimal two-sided ideals, in one-one correspondence with the partitions of k. Each of these minimal ideals, corresponding to a partition P, is an algebra of all g_P by g_P-matrices over F, where g_P is determined by P so that

$$\sum_P g_P^2 = k!$$

A primitive idempotent e_P in the ideal corresponding to P is constructed by normalizing a "Young symmetrizer" as in (3.8) of [28], p. 126. (Obviously e_P is not unique if $g_P > 1$.) The algebra $S^k(A)$ is Weyl's algebra A_k of "bisymmetric transformations" ([28], p. 127).

Letting $F[S_k]$ operate from the left on $\otimes^k \mathcal{V}$ and $\otimes^k A$ from the right, as before, it follows from the above that if $\mathcal{W}$ _is an irreducible S_k-submodule of $\otimes^k \mathcal{V}$ not annihilated by the ideal in $F[S_k]$ corresponding to P, then $e_P(\otimes^k \mathcal{V})$ intersects $\mathcal{W}$ in a one-dimensional subspace_. With $\mathcal{U}$ as above, corresponding to P, we then have $e_P(\otimes^k \mathcal{V}) = e_P \mathcal{U} = e_P \mathcal{U}_1 \oplus \dots \oplus e_P \mathcal{U}_r$, each $e_P \mathcal{U}_i$ having dimension one. The generators u_i may be taken in $e_P \mathcal{U}_i$, so that $e_P u_i = u_i$, $1 \le i \le r$.

These u_i are a basis for $e_P(\otimes^k \mathcal{V})$, a subspace of $\otimes^k \mathcal{V}$ stable under $S^k(A)$, and $S^k(A)$ induces in $e_P(\otimes^k \mathcal{V})$ the full algebra of F-endomorphisms of this space. If $e_P^{(1)}, \dots, e_P^{(g_P)}$ is a complete set of orthogonal primitive idempotents in the minimal ideal of $F[S_k]$ corresponding to P, then these are connected by matrix-units in this ideal, say $e_{ij} e_P^{(j)} = e_P^{(i)}$, $1 \le i, j \le g_P$, so that $e_P^{(i)}(\otimes^k \mathcal{V})$ and $e_P^{(j)}(\otimes^k \mathcal{V})$ are isomorphic $S^k(A)$-modules. The sum, over i, of these $e_P^{(i)}(\otimes^k \mathcal{V})$ is equal to $\sum_{i=1}^{g_P} e_P^{(i)} \mathcal{U} = \mathcal{U}$. That is, $\mathcal{U}$ is a sum of irreducible $S^k(A)$-_modules, all of them isomorphic to_ $e_P(\otimes^k \mathcal{V})$.

Every $S^k(A)$-submodule of $\otimes^k \mathcal{V}$ isomorphic to a simple submodule of $\mathcal{U}$ is contained in $\mathcal{U}$; for if $\mathcal{W}$ is a simple submodule isomorphic to a submodule of $\mathcal{U}$, there is an automorphism of the $S^k(A)$-module $\otimes^k \mathcal{V}$ mapping a submodule of $\mathcal{U}$ onto $\mathcal{W}$. But all $S^k(A)$-endomorphisms of $\otimes^k \mathcal{V}$ are of the form $w \mapsto sw$, $s \in F[S_k]$, and therefore map $\mathcal{U}$ into $\mathcal{U}$; hence $\mathcal{W} \subseteq \mathcal{U}$. This is the classical result that _the S_k-isotypic and $S^k(A)$-isotypic constituents of $\otimes^k V$ coincide; there are as many of them as partitions P of k with $e_P(\otimes^k V) \neq 0$ for one corresponding_ (normalized) _Young symmetrizer_ e_P.

In Lemma (4.4.B) of [28], Weyl determines those P for which the last condition is satisfied: The Young diagram that defines e_P has at most k (non-empty) rows. If the number of rows exceeds the dimension of V, then $e_P(\otimes^k V) = 0$; otherwise $e_P(\otimes^k V) \neq 0$. In applications of this theory, we shall mainly be interested in cases where $k \leq d$. For these, we have:

If $k \leq d$, $S^k(A)$ is a direct sum of minimal two-sided ideals, each of which is a full matrix algebra over F. There are $p(k)$ of these, in number corresponding to the partitions P of k. Their dimensions are d_P^2, where d_P is the multiplicity in $\otimes^k V$ of the irreducible representation of S_k corresponding to the partition P. In the action of $S^k(A)$ on $\otimes^k V$, the multiplicity of the irreducible representation of $S^k(A)$ corresponding to P is g_P, where g_P^2 is the dimension of the corresponding minimal ideal in $F[S_k]$.

2. A SPECIAL MINIMAL IDEAL IN $S^k(A)$.

As in §1, let A be at first a matrix algebra $M_d(F)$, acting by right multiplication on the space $V = F^d_{(row)}$ of d-tuples from F, written as row-vectors. Then $A = \mathrm{End}_F(V)$. The d-rowed column vectors from F may be identified in the canonical way with the dual space V^* of V, and these constitute a right module for the opposite algebra A^{op} when we define $(\xi_i)(\alpha_{ij}) = (\alpha_{ij})(\xi_i)$, the latter being ordinary matrix multiplication; here $(\xi_i) \in V^*$ and (α_{ij}) is a $d \times d$ matrix. Then $M_d(F)^{op}$ is the F-endomorphism algebra (written on the right!) of V^*.

If V is a vector space over F, we use the notation $\Lambda^k(V)$ for both the space of skew-symmetric tensors in $\otimes^k V$ and for the subspace of the exterior algebra $\Lambda(V)$ that usually bears this designation. An F-isomorphism of the two spaces is obtained by restricting to the former the canonical map of $\otimes^k V$ onto the latter, _i.e._, the map sending $v_1 \otimes \cdots \otimes v_k$ to $v_1 \wedge \cdots \wedge v_k \in \Lambda(V)$. Clearly $\Lambda^k(V)$ ($\subseteq \otimes^k V$) is stable under $S^k(A) = S^k(\mathrm{End}_F(V))$. The action of $S^k(A)$ on $\Lambda^k(V)$ is irreducible, even absolutely irreducible for $1 \leq k \leq d$ ([15], Chapter 7, §6). Acting in $\Lambda^k(V)$ ($\subseteq \Lambda(V)$), $\sum_{\pi \in S_k} a_{\pi(1)} \otimes \cdots \otimes a_{\pi(k)}$, $a_i \in A$, sends $v_1 \wedge \cdots \wedge v_k$ to $\sum_{\pi \in S_k} v_1 a_{\pi(1)} \wedge \cdots \wedge v_k a_{\pi(k)}$.

Now $\Lambda^j(V)$ and $\Lambda^j(V^*)$ are dual spaces in a pairing sending

$v_1 \wedge \dots \wedge v_j$, $v_i \in V$, and $v_1^* \wedge \dots \wedge v_j^*$, $v_i^* \in V^*$, to the scalar

$$\sum_{\pi \in S_j} \operatorname{sgn}(\pi) < v_{\pi(1)}, v_1^* > \dots < v_{\pi(j)}, v_j^* >.$$

Also, $\Lambda^j(V^*)$ and $\Lambda^{d-j}(V^*)$ are dual spaces in a pairing sending $v_1^* \wedge \dots \wedge v_j^*$ and $w_1^* \wedge \dots \wedge w_{d-j}^*$ to the scalar Δ such that $v_1^* \wedge \dots \wedge v_j^* \wedge w_1^* \wedge \dots \wedge w_{d-j}^* \in \Lambda^d(V^*)$ is equal to the multiple by Δ of a fixed non-zero element Ω in $\Lambda^d(V^*)$, a one-dimensional space. It follows that we have an isomorphism of vector spaces between $\Lambda^j(V)$ and $\Lambda^{d-j}(V^*)$, resulting from these identifications with the dual space of $\Lambda^j(V^*)$. We use the isomorphism to transport the structure of $S^j(A)$-module from $\Lambda^j(V)$ to $\Lambda^{d-j}(V^*)$, and relate this structure to that on $\Lambda^{d-j}(V^*)$ as $S^{d-j}(A^{op})$-module.

First let $a \in A$ have simple eigenvalues $\gamma_1, \dots, \gamma_d$, all in F, and let $e_1, \dots, e_d$ be a basis for V, $e_i a = \gamma_i e_i$, $1 \le i \le d$. Let $f_1, \dots, f_d$ be the corresponding dual basis for V^*. We again write "a" for the image of a under the anti-isomorphism of A onto A^{op} which is the identity on the underlying set. Thinking of A as matrices, V as row-vectors and V^* as column vectors, we then have, for a as above,

$$< e_r, f_s \cdot a > = e_r(af_s) = (e_r a)f_s = \gamma_r e_r f_s = \gamma_r < e_r, f_s >$$
$$= \gamma_r \delta_{r,s}, \ 1 \le r, s \le d,$$

so that $f_s \cdot a = \gamma_s f_s$, $1 \le s \le d$. We fix our Ω above as $\Omega = \mu f_1 \wedge \dots \wedge f_d$, $0 \ne \mu \in F$, and we let $\rho_j(a) = a \otimes 1 \otimes \dots \otimes 1 + \dots + 1 \otimes \dots \otimes 1 \otimes a \in S^j(A)$, as in §III.2.

For $i_1 < i_2 < \dots < i_j$, $\rho_j(a)$ sends $e_{i_1} \wedge \dots \wedge e_{i_j}$ to $(\gamma_{i_1} + \dots + \gamma_{i_j}) e_{i_1} \wedge \dots \wedge e_{i_j}$. The corresponding action of the transpose of $\rho_j(a)$ on the dual space of $\Lambda^j(V)$, identified with $\Lambda^j(V^*)$ sends $f_{\nu_1} \wedge \dots \wedge f_{\nu_j}$ $(\nu_1 < \nu_2 < \dots < \nu_j)$ to

$$\delta_{i_1,\nu_1} \delta_{i_2,\nu_2} \cdots \delta_{i_j,\nu_j} (\gamma_{i_1} + \dots + \gamma_{i_j}) f_{\nu_1} \wedge \dots \wedge f_{\nu_j} .$$

In the dual pairing of $\Lambda^j(V^*)$ and $\Lambda^{d-j}(V^*)$, the basis for $\Lambda^{d-j}(V^*)$ dual to the $f_{\nu_1} \wedge \dots \wedge f_{\nu_j}$ is the set $\{\mu^{-1} \varepsilon_{\mu_1,\dots,\mu_{d-j}} f_{\mu_1} \wedge \dots \wedge f_{\mu_{d-j}}\}$, where $\mu_1 < \mu_2 < \dots < \mu_{d-j}$, and where if $\{1, \dots, d\} = \{\nu_1, \dots, \nu_j\} \cup \{\mu_1, \dots, \mu_{d-j}\}$, $\varepsilon_{\mu_1,\dots,\mu_{d-j}}$ is the sign of the permutation sending 1 to ν_1, 2 to $\nu_2, \dots,$ j to ν_j,

$j+1$ to $\mu_1,\ldots,$ d to μ_{d-j}. In the identification of $\Lambda^j(V)$ and $\Lambda^{d-j}(V^*)$ as the dual space of $\Lambda^j(V^*)$, the element of $\Lambda^{d-j}(V^*)$ corresponding to the basic element $e_{i_1} \wedge \ldots \wedge e_{i_j}$ of $\Lambda^j(V)$ is thus

$$\mu^{-1}\ \varepsilon_{\mu_1,\ldots,\,\mu_{d-j}}\ f_{\mu_1} \wedge \ldots \wedge f_{\mu_{d-j}}\ , \text{ where}$$

$$\{\mu_1,\ldots,\mu_{d-j}\} \cup \{i_1,\ldots,i_j\} = \{1,\ldots,d\}\ .$$

Denote by $\sigma_{d-j}(a)$ the action of $\rho_j(a)$ on $\Lambda^{d-j}(V^*)$, transported by our isomorphism from the above action on $\Lambda^j(V)$. Then $\sigma_{d-j}(a)$, for the diagonal a above, has the element at the end of the last paragraph as eigenvector, with eigenvalue $\gamma_{i_1}+\ldots+\gamma_{i_j}$. That is, $\sigma_{d-j}(a)$ sends $f_{\mu_1} \wedge \ldots \wedge f_{\mu_{d-j}}$ to $(t(a) - (\gamma_{\mu_1}+\ldots+\gamma_{\mu_{d-1}}))f_{\mu_1} \wedge \ldots \wedge f_{\mu_{d-j}}$, where $t(a)$ is the trace of a.

When we regard A^{op} as the F-endomorphisms of V^*, we may consider $\rho_{d-j}(a) = a \otimes 1 \otimes\ldots\otimes 1 +\ldots+ 1 \otimes \ldots \otimes 1 \otimes a \in S^{d-j}(A^{op})$ as operating (still on the right) on $\Lambda^{d-j}(V^*)$. Recalling that the right action of $a \in A^{op}$ on $v^* \in V^*$ is just the action of the transpose of a, when a is regarded as an element of A, we see that the action of $\rho_{d-j}(a) \in S^{d-j}(A^{op})$ for our diagonal a sends

$$f_{\mu_1} \wedge \ldots \wedge f_{\mu_{d-j}} \quad \text{to} \quad (\gamma_{\mu_1}+\ldots+\gamma_{\mu_{d-j}})f_{\mu_1} \wedge \ldots \wedge f_{\mu_{d-j}}\ .$$

That is, whenever $a \in A = \mathrm{End}_F(V)$ is diagonalizable, the transported action $\sigma_{d-j}(a)$ on $\Lambda^{d-j}(V^*)$ and the action of $\rho_{d-j}(a) \in S^{d-j}(A)^{op})$ on $\Lambda^{d-j}(V^*)$ are related by

(1) $$w(\rho_{d-j}(a) + \sigma_{d-j}(a)) = t(a)w$$

for all $w \in \Lambda^{d-j}(V^*)$.

But one sees easily that A is spanned by matrices diagonalizable in F, from which it follows by the linearity in a of both sides of (1) that (1) holds for all $w \in \Lambda^{d-j}(V^*)$ and all $a \in \mathrm{End}_F(V)$.

Since ρ_j satisfies the (j+1)-th symmetric identity of §III.2, so does σ_{d-j}. With $k = d-j$, this means that the mapping φ_k sending $a \in A$ (= A^{op}, as sets) to the action on $\Lambda^k(V^*)$ of $\rho_k(a) \in S^k(A^{op})$ has the property that φ_k satisfies the (k+1)-th symmetric identity, while σ_k: $\sigma_k(a) = t(a)\,1 - \varphi_k(a)$, satisfies the (d-k+1)-th identity.

Let M be the ideal in $S^k(A^{op})$ generated by all elements of the form

$$g(a) = \sum_{P\leftrightarrow(s_1,\dots,s_{d-k+1})} \operatorname{sgn}(C(P))\,|C(P)|\,(t(a)-\rho_k(a))^{s_1}(t(a^2)-\rho_k(a^2))^{s_2}\cdots \tag{2}$$
$$(t(a^{d-k+1}) - \rho_k(a^{d-k+1}))^{s_{d-k+1}}$$

Evidently M lies in the kernel of the representation of $S^k(A^{op})$ on $\Lambda^k(V^*)$, by our remarks of the last paragraph. We show below: If $u \in \otimes^k(V^*)$ is annihilated by M, then $u \in \Lambda^k(V^*)$. Thus M is precisely the annihilator of $\Lambda^k(V^*)$ in the subalgebra $S^k(A^{op})$ of endomorphisms of $\otimes^k V^*$. Namely, if M is not a maximal ideal in $S^k(A^{op})$ there is an ideal of the form $B_1 \oplus B_2$, where B_1 and B_2 are non-zero, that intersects M only in zero. Then M annihilates two non-isomorphic irreducible $S^k(A^{op})$-modules, and therefore two non-isomorphic submodules of $\otimes^k V^*$. This would force $\Lambda^k(V^*)$ to be reducible as $S^k(A^{op})$-module; but $\Lambda^k(V^*)$ is irreducible for $1 \le k \le d$. (For example, see Theorem VII.6 of [15].) Being maximal and contained in the annihilator of $\Lambda^k(V^*)$, M must coincide with this annihilator.

To prove the assertion that the submodule of $\otimes^k(V^*)$ annihilated by M is $\Lambda^k(V^*)$, we may assume $V^* = F^d$, $A^{op} = M_d(F)$. We think of F^d as row-vectors and of $M_d(F)$ multiplying these from the right. Letting $e_1,\dots, e_d$ be the standard basis for F^d, we let

$$u = \sum \beta_{i_1,\dots,i_k}\, e_{i_1} \otimes \cdots \otimes e_{i_k} \in \otimes^k(F^d)$$

be annihilated by all $g(a)$, $a \in M_d(F)$; the sum is over distinct sequences of k indices, with repeated indices allowed.

First note that $\beta_{i_1,\dots,i_k} = 0$ if two indices are equal. For if in a given term some index, say 1, is repeated, then another index, say d, must be omitted. Then take $a = E_{dd} - E_{11}$, in the usual notation for matrix units. We determine the effect of $g(a)$ on $e_{i_1} \otimes \cdots \otimes e_{i_k}$: If j is odd, $a^j = a$, $t(a^j) = 0$, and $t(a) - \rho_k(a) = -\rho_k(a)$. If j is even, $a^j = E_{dd} + E_{11}$, $t(a^j) - \rho_k(a^j) = 2 - \rho_k(E_{11} + E_{dd})$. If exactly m of the indices $i_1,\dots, i_k$ are equal to 1 (and none to d), $-\rho_k(a)$ sends $e_{i_1} \otimes \cdots \otimes e_{i_k}$ to $m\,(e_{i_1} \otimes \cdots \otimes e_{i_k})$, while $t(a^2) - \rho_k(a^2)$ sends this element to $(2-m)\,(e_{i_1} \otimes \cdots \otimes e_{i_k})$. Thus $g(a)$ sends this basic tensor to

$$\sum_P \operatorname{sgn}(C(P))\,|C(P)|\, m^{s_1+s_3+\cdots}\,(2-m)^{s_2+s_4+\cdots} \tag{3}$$

times itself.

The permutations in $C(P)$ are even or odd according as $s_2 + s_4 + \cdots$ is even or odd. Thus (3) is equal to

$$\sum_P |C(P)|\, m^{s_1+s_3+\cdots}\,(m-2)^{s_2+s_4+\cdots},$$

evidently a positive integer since $m \geq 2$ and since the term with $P \leftrightarrow (d-k+1,0,\ldots,0)$ is $m^{d-k+1} > 0$. Moreover, every $e_{j_1} \otimes \ldots \otimes e_{j_k}$ is an eigenvector for $g(a)$, with our a. It follows that, since the eigenvalue for the term $e_{i_1} \otimes \ldots \otimes e_{i_k}$ is non-zero, its coefficient must be zero.

We may now assume that all basic tensors $e_{i_1} \otimes \ldots \otimes e_{i_k}$ actually present in u have distinct indices. To see that u is skew, it suffices to show that a transposition of a pair of indices corresponds to terms whose coefficients are negatives of one another. Thus, without loss of generality, it will be enough to show $\beta_{2,1,3,4,\ldots,k} + \beta_{1,2,3,4,\ldots,k} = 0$. Here we take $a = E_{12} + E_{21}$, so that $t(a^j) - \rho_k(a^j) = -\rho_k(a^j)$ for odd j. The action of this on four basic tensors in $\otimes^k(F^d)$ is as follows:

$e_1 \otimes e_1 \otimes e_3 \otimes \ldots \otimes e_k \to -(e_1 \otimes e_2 \otimes e_3 \otimes \ldots \otimes e_k + e_2 \otimes e_1 \otimes e_3 \otimes \ldots \otimes e_k);$

$e_2 \otimes e_2 \otimes e_3 \otimes \ldots \otimes e_k \to$ (above)

$e_1 \otimes e_2 \otimes e_3 \otimes \ldots \otimes e_k \to -(e_1 \otimes e_1 \otimes e_3 \otimes \ldots \otimes e_k + e_2 \otimes e_2 \otimes \ldots \otimes e_k);$

$e_2 \otimes e_1 \otimes e_3 \otimes \ldots \otimes e_k \to$ (above)

Meanwhile, if j is even, $t(a^j) - \rho_k(a^j) = 2 - \rho_k(E_{11} + E_{22})$ annihilates all four of the above.

If $e_{j_1} \otimes e_{j_2} \otimes \ldots \otimes e_{j_k}$ is any other basic tensor with distinct indices, and if some $j_\nu > k$, the image of this basis element under $g(a)$ will still be a combination (possibly zero) of basic tensors involving this new index. Likewise if one of j_1, j_2 is greater than 2, our basic tensor will be sent into a combination of basic tensors with the same property. Thus the only occurrences of tensors which are combinations of our four in the image of the action of $g(a)$ on u are in the terms arising as the image of

(4) $$\beta_{1,2,3,\ldots,k}\, e_1 \otimes e_2 \otimes \ldots \otimes e_k + \beta_{2,1,3,\ldots,k}\, e_2 \otimes e_1 \otimes e_3 \otimes \ldots \otimes e_k .$$

Now we have seen that the effect of $g(a)$ on (4) is the same as that of

(5) $$\sum_{P \leftrightarrow (s_1,0,s_3,0,\ldots)} |C(P)|\, (-\rho_k(a))^{s_1+s_3+\cdots}$$

The terms of (5) with $s_1 + s_3 + \ldots$ even have sum sending both $e_1 \otimes e_2 \otimes \ldots \otimes e_k$ and $e_2 \otimes e_1 \otimes \ldots \otimes e_k$ to

$$v = \left(\sum_{\substack{P \leftrightarrow (s_1,0,s_3,0,\dots) \\ s_1+s_3+\dots \ \text{even}}} |C(P)| \ 2^{\frac{s_1+s_3+\dots}{2}} \right) \cdot (e_1 \otimes e_2 \otimes \dots \otimes e_k$$

$$+ e_2 \otimes e_1 \otimes \dots \otimes e_k) ,$$

while the remaining terms combine to send both of these basic tensors to

$$w = \left(\sum_{\substack{P \leftrightarrow (s_1,0,s_3,0,\dots) \\ s_1+s_3+\dots \ \text{odd}}} -|C(P)| \ 2^{\frac{(s_1+s_3+\dots)+1}{2}} \right) (e_1 \otimes e_1 \otimes \dots \otimes e_k$$

$$+ e_2 \otimes e_2 \otimes \dots \otimes e_k).$$

Thus the terms in $ug(a)$ arising from the terms in the two basic tensors above are

$$(\beta_{1,2,\dots,k} + \beta_{2,1,\dots,k})(v+w)$$

and no basic tensor among our fundamental four occurs elsewhere in $ug(a)$. Since v and w are not both zero (consider $P \leftrightarrow (d-k+1,0,\dots,0)$), and are linearly independent if neither is zero, we have

$$\beta_{1,2,\dots,k} + \beta_{2,1,3,\dots,k} = 0, \quad \text{and} \quad u \ \text{is } \underline{\text{skew}}.$$

It follows from the above, combined with §3 of Chapter III, that:

PROPOSITION IV.1: If $A = \text{End}_F(V)$, $[V:F] = d$, and if $1 \le k < d$, there is an associative algebra B and a linear mapping $\rho: A \to B$, such that:

i) ρ is a Lie morphism;

ii) $\rho(1) = k \cdot 1$;

iii) ρ satisfies the $(k+1)$-th symmetric identity;

iv) $\sigma: A \to B$, defined by $\sigma(a) = t(a)\,1 - \rho(a)$, satisfies the $(d-k+1)$-th symmetric identity;

v) B and ρ are universal for i) - iv).

Namely, to obtain ρ we combine $\rho_k: A \to S^k(A)$ with projection onto the quotient B of $S^k(A)$ by the ideal generated by all $g(a)$ as in (2). Since the pair $(S^k(A), \rho_k)$ is universal for i) - iii) by Proposition III.3, it follows that (B,ρ) is universal for i) - iv).

We have seen above that B acts irreducibly in $\Lambda^k(V)$, since $S^k(A)$ does, and the same holds upon arbitrary extension of the base field. Thus B is a central simple algebra over F, $[B:F] = \binom{d}{k}^2$.

3. SYMMETRIC POWERS FOR CENTRAL SIMPLE INVOLUTORIAL ALGEBRAS; THE SPLIT CASE.

By a central simple involutorial algebra over F we understand an associative algebra A with involution $*$ over F, having no proper $*$-stable ideals, and such that F is the set of $*$-fixed elements of the center of A. It is well known that either A is a central simple algebra over F or A is a central simple algebra over its center Z, a quadratic extension of F, or A is the direct sum of two minimal ideals, interchanged by the involution, each of which is a central simple algebra over F. The last two cases are those "of second kind". By passing to a (finite) field extension K, the second case becomes the third, and in the third the minimal ideals become matrix algebras $M_d(K)$.

The involution will not enter into further consideration for some time. When the involution is ignored, we have already examined $S^k(A)$ in the split case where A is central simple. We now study $S^k(A)$ in the split case of second kind, where $A \cong M_d(F) \oplus M_d(F)^{op}$. From another point of view, $A \cong \mathrm{End}_F(V) \oplus \mathrm{End}_F(V^*)$, where here V^* is the dual space of V, and A may be regarded as the subalgebra of $\mathrm{End}_F(V \oplus V^*)$ stabilizing both V and V^*. Thus

$$R = \otimes^k A \subseteq \otimes^k(\mathrm{End}_F(V \oplus V^*)) = \mathrm{End}_F(\otimes^k(V \oplus V^*))$$

is a semisimple algebra of endomorphisms of $\otimes^k(V \oplus V^*)$. Likewise the symmetric group S_k acts in $\otimes^k(V \oplus V^*)$, and the centralizer of S_k in R is $S^k(A)$ (cf. §1). It follows by general principles that $S^k(A)$ is a semi-simple algebra of endomorphisms of $\otimes^k(V \oplus V^*)$.

If e_1 and e_2 are the respective projections of $1 \in A$ on the summands $\mathrm{End}(V)$, $\mathrm{End}(V^*)$, these are the units of these algebras and are a complete set of central idempotents in A. When $r + s = k$, let $e_{r,s} \in S^k(A)$ be the element obtained by symmetrizing

$$\overbrace{e_1 \otimes \cdots \otimes e_1}^{r} \otimes \overbrace{e_2 \otimes \cdots \otimes e_2}^{s},$$

without repetitions. Thus $e_{r,s}$ is the sum of $\binom{k}{r}$ terms of the form $u_1 \otimes \cdots \otimes u_k$, where exactly r of the u_i's are equal to e_1, and the rest are e_2. Then $e_{r,s}$ is a central idempotent in $S^k(A)$, $e_{r,s}e_{r',s'} = 0$ if $r \neq r'$, and $\sum_{r=0}^{k} e_{r,s} = 1 \otimes \cdots \otimes 1$.

Accordingly there is a direct decomposition of $S^k(A)$ into $k + 1$ ideals $S^k(A)e_{r,s}$. From the linear generation of $S^k(A)$ by elements $a \otimes a \otimes \cdots \otimes a$, $a \in A$, we see that $S^k(A)e_{r,s}$ is spanned by the elements obtained by symmetrizing all

$$(6) \qquad \overbrace{ae_1 \otimes \dots \otimes ae_1}^{r} \otimes \overbrace{ae_2 \otimes \dots \otimes ae_2}^{s}, \quad a \in A .$$

Consider the mapping of $\otimes^k A$ into $(\otimes^r Ae_1) \otimes (\otimes^s Ae_2)$ sending $u_1 \otimes \dots \otimes u_k$ to

$$u_1 e_1 \otimes \dots \otimes u_r e_1 \otimes u_{r+1} e_2 \otimes \dots \otimes u_k e_2 .$$

Call this mapping $\varphi_{r,s}$; it is a homomorphism of algebras with unit, and we see from the remarks preceding (6) above that $\varphi_{r,s}$ maps $S^k(A)$ into $S^r(Ae_1) \otimes S^s(Ae_2)$, with $\varphi_{r,s}(S^k(A)e_{r',s'}) = 0$ if $r \neq r'$. The element obtained by symmetrizing (6) is sent to (6), so the image of $S^k(A)$ under $\varphi_{r,s}$ is spanned by the elements (6) (no longer symmetrized!). Writing $a = a_1 + a_2$, where $a_i \in A\,e_i$, $i = 1,2$, the element (6) has the form

$$a_1 e_1 \otimes \dots \otimes a_1 e_1 \otimes a_2 e_2 \otimes \dots \otimes a_2 e_2$$

But elements of the form $\overbrace{a_i e_i \otimes \dots \otimes a_i e_i}^{t}$, $i = 1,2$, span $S^t(Ae_i)$; that is, $\varphi_{r,s}$ <u>maps</u> $S^k(A)e_{r,s}$ <u>onto</u> $S^r(Ae_1) \otimes S^s(Ae_2)$.

In fact $\varphi_{r,s}$ establishes an isomorphism between $S^k(A)e_{r,s}$ and $S^r(Ae_1) \otimes S^s(Ae_2)$. One can either see this by noting that symmetrizing a general element u of $S^r(Ae_1) \otimes S^s(Ae_2) \subseteq \otimes^k A$ gives an element of $S^k(A)e_{r,s}$ which maps to u under $\varphi_{r,s}$, and that the symmetrized elements resulting span $S^k(A)e_{r,s}$, or by use of the identity

$$(7) \qquad \binom{m+n+k-1}{k} = \sum_{r=0}^{k} \binom{m+r-1}{r} \binom{m+k-r-1}{k-r}$$

(<u>cf</u>. W. Feller, <u>An Introduction to Probability Theory and its Applications</u> (1st ed.), ex. 11, p. 48 [Wiley, New York, 1950].) The left side of (7), for $m = n = d$, is the dimension of $S^k(A)$, while the right side is the sum of the dimensions of the $S^r(Ae_1) \otimes S^{k-r}(Ae_2)$, $0 \leq r \leq k$. Since the mappings of the last paragraph give a homomorphism of $S^k(A)$ <u>onto</u> $\sum_r S^r(Ae_1) \otimes S^{k-r}(Ae_2)$, equality of dimensions shows that the map is an isomorphism. It follows that each $\varphi_{r,s}$ is an isomorphism as claimed.

The structure of $S^k(A)$ is now established: $S^k(A)$ is the direct sum of the ideals $S^k(A)e_{r,s}$ $(r + s = k)$, which in turn are isomorphic to $S^r(Ae_1) \otimes S^s(Ae_2)$. Assuming $k \leq d$, $S^r(Ae_1)$ is the sum of $p(r)$ minimal ideals, each of which is a split central simple algebra, as in §1, and likewise for $S^s(Ae_2)$. Thus $S^k(A)$ is a sum of

$$\sum_{r=0}^{k} p(r)p(k-r)$$

minimal ideals, each of which is a split central simple algebra.

Here it is worthwhile noting the action of $\rho_k(e_1-e_2) \in S^k(A)$ on $S^k(A)$. This is a central element, whose product with each of the terms $u_1 \otimes \ldots \otimes u_k$ ($u_j = e_1$ or e_2) in $e_{r,s}$ is equal to $(r-s)\, u_1 \otimes \ldots \otimes u_k$. Thus $e_{r,s}\, \rho_k(e_1-e_2) = (r-s)e_{r,s}$. That is, $\rho_k(e_1-e_2)$ _acts diagonally on_ $S^k(A)$, _with eigenvalues_ k, $k-2$, $k-4, \ldots, -k+2, -k$; _the subspace belonging to the eigenvalue_ $r-s$ $(r+s = k)$ is $S^k(A)e_{r,s} \cong S^r(Ae_1) \otimes S^s(Ae_2)$.

4. THE GENERAL CASE.

In general, if A is central simple over F, $[A:F] = d^2$, if $k \le d$, and if K is an extension field of F splitting A (if needed, $[K:F]$ may be assumed finite), we have $S^k(A)_K = S^k(A_K) = S^k(M_d(K))$, a semisimple algebra over K with structure as in §1. It follows that $S^k(A)$ is semisimple, with center of dimension equal to $p(k)$, and that $S^k(A)$ becomes the sum of $p(k)$ central simple ideals on extension of the base field to K. Thus $S^k(A)$ is the sum of at most $p(k)$ minimal ideals.

In particular, these considerations apply if A is a central simple involutorial algebra of second kind in the sense of §3, where the center Z of A (as algebra only) is a quadratic extension field of F. If we ignore the involution and regard A simply as a central simple Z-algebra, we denote this structure by $A(Z)$, and work over Z as ground field. With $d^2 = [A(Z):Z]$ and $k \le d$, the last paragraph applies to $S^k(A(Z))$.

When the algebra A over F (rather than $A(Z)$, over Z) is considered, we have $Z = F(\zeta)$, $\zeta^* = -\zeta$, $\zeta^2 = \gamma \in F$, $\gamma \notin F^2$. A splitting field K for A is chosen to contain an element λ such that $\lambda^2 = \gamma$, leading to the split setting at the end of the first paragraph of §3. In $A_K = A \otimes K$, the central idempotents e_1 and e_2 are fixed as $\frac{1}{2}(1 \otimes 1 \pm \gamma^{-1}\zeta \otimes \lambda)$, respectively. Then the structural information of §3 on $S^k(A)_K = S^k(A_K)$ applies whenever $k \le d$.

In A_K, $e_1 - e_2 = \gamma^{-1}\zeta \otimes \lambda$, so that in $S^k(A)_K$, $\rho_k(e_1-e_2) = \gamma^{-1}(\rho_k(\zeta) \otimes \lambda)$ acts diagonally with integral eigenvalues k, $k-2, \ldots, -k$ as in §3. Thus $\rho_k(e_1-e_2)^2 = \gamma^{-2}(\rho_k(\zeta)^2 \otimes \lambda^2) = \gamma^{-1}(\rho_k(\zeta)^2 \otimes 1)$ acts diagonally with non-negative integral eigenvalues k^2, $(k-2)^2, \ldots$. That is, $\rho_k(\zeta) \in S^k(A)$ satisfies the polynomial $(X^2-k^2\gamma)(X^2-(k-2)^2\gamma)\ldots(X^2-\gamma)$ if k is _odd_, and the polynomial

$$(X^2-k^2\gamma)(X^2-(k-2)^2\gamma)\ \ldots\ (X^2-4\gamma)X$$

if k is _even_.

Thus $S^k(A)$ is the direct sum of the ideals annihilated by $\rho_k(\zeta)^2 - j^2\gamma$, $j = 1,3,\ldots,k$, if k is odd, and of the ideals annihilated by $\rho_k(\zeta)$ and by $\rho_k(\zeta)^2 - j^2\gamma$ for $j = 2,4,\ldots,k$, if k is even. Our analysis shows that these ideals become

$$S^r(A_K e_1) \otimes_K S^s(A_K e_2) \oplus S^s(A_K e_1) \otimes_K S^r(A_K e_2)$$

for $0 \le r \le [\frac{k-1}{2}]$, $r + s = k$, $s - r = j$, and when k is <u>even</u>, the ideal annihilated by $\rho_k(\zeta)$ becomes

$$S^{\frac{k}{2}}(A_K e_1) \otimes S^{\frac{k}{2}}(A_K e_2)$$

In particular, the action of $\rho_k(\zeta)$ on any irreducible $S^k(A)$-module must satisfy $X^2 - (k-2r)^2 \gamma = 0$, $0 \le r \le [\frac{k}{2}]$; furthermore, when $2r = k$, $\rho_k(\zeta)$ must annihilate the irreducible module.

Returning to the case where A is central simple over F, $[A:F] = d^2$, we study the form that the situation of §2 takes on. Here for $a \in A$, the matrix trace has to be replaced by the <u>reduced trace</u> $t(a)$. Because we are dealing with characteristic zero only, we may summarize the properties of this function by saying that every a is uniquely representable as

$$a = \frac{1}{d} t(a) 1 + a_0, \quad a_0 \in [AA],$$

so that $t(1) = d$, $t(ab) = t(ba)$. The ideal in $S^k(A)$ $(1 < k < d)$ generated by all elements

$$(8) \qquad \sum_{P \leftrightarrow (s_1,\ldots,s_{d-k+1})} \operatorname{sgn}(C(P)) \, |C(P)| \sigma_k(a)^{s_1} \sigma_k(a^2)^{s_2} \ldots \sigma_k(a^{d-k+1})^{s_{d-k+1}},$$

where $\sigma_k(a) = t(a) 1 - \rho_k(a)$, contains a basis for the ideal in $S^k(A)_K = S^k(A_K)$ generated by elements (8) with A replaced by A_K (K a splitting field-one sees this by polarization). We have seen in §2 that this ideal is maximal in $S^k(A_K)$ and that the quotient is the matrix algebra $M_{\binom{d}{k}}(K)$. It follows that the ideal generated by the elements (8) in $S^k(A)$ is also maximal, with quotient algebra a central simple algebra of degree $\binom{d}{k}$. That is, the conclusions and remarks of Proposition 1 of §2 remain valid in the general case. We formulate these as:

<u>PROPOSITION IV.2</u>: <u>Let A be a central simple algebra over F, $[A:F] = d^2$, and let $1 \le k < d$. Then there is an associative algebra B and a linear mapping $\rho: A \to B$, such that</u>

i) ρ <u>is a Lie morphism;</u>

ii) $\rho(1) = k1$;

iii) ρ <u>satisfies the $(k+1)$-th symmetric identity;</u>

iv) $\sigma = t - \rho$ <u>satisfies the $(d-k+1)$-th symmetric identity;</u>

v) B <u>and</u> ρ <u>are universal for i) - iv).</u>

B is central simple over F, with $[B:F] = \binom{d}{k}^2$. Namely, B is the quotient of $S^k(A)$ by the ideal generated by the elements (8), and ρ is the composite of ρ_k with the canonical homomorphism $S^k(A) \to B$.

Finally, we investigate the presence of abelian ideals in $S^k(A)$, $k \le d$. For $k = 0$, $S^o(A) = F$ is abelian, so we assume A is central simple, $1 \le k \le d$. First consider the split case $A = \mathrm{End}_F(V)$, $[V:F] = d$; in this case, as we have seen in §1, the minimal ideals in $S^k(A)$ are the F-endomorphisms of vector spaces $e_p(\otimes^k V)$, where e_p is a primitive idempotent constructed from a Young symmetrizer in $F[S_k]$.

Following the notation of Chapter IV of [28], this means that e_p is a non-zero scalar multiple of an element c of $F[S_k]$ associated with a Young diagram T, as follows: Let R(T) be the subgroup of S_k stabilizing the rows of T, C(T) the subgroup stabilizing the columns. Then each element of the set C(T)R(T) has a unique representation as $\sigma\rho$, $\sigma \in C(T)$, $\rho \in R(T)$. The element $c = c_T$ is defined by $c = \sum_{\substack{\sigma \in C(T) \\ \rho \in R(T)}} \mathrm{sgn}(\sigma)\sigma\rho \in F[S_k]$.

If $S^k(A)$ is to have abelian ideals, there must be a Young diagram T for which $c_T(\otimes^k V)$ has dimension 1. <u>This is not the case if</u> $1 \le k \le d$; for then if $i_1 < \dots < i_k$ are distinct indices and $e_1, \dots, e_d$ is a fixed basis for V, it is clear that $c(e_{i_1} \otimes \dots \otimes e_{i_k})$ is a linear combination of the elements $e_{i_{\pi(1)}} \otimes \dots \otimes e_{i_{\pi(k)}}$, $\pi \in C(T)R(T) \subseteq S_k$ with the coefficient of $e_{i_1} \otimes \dots \otimes e_{i_k}$ being 1. Taking two distinct subsets $i_1, \dots, i_k$, we thus obtain linearly independent elements of $c_T(\otimes^k V)$.

Next suppose $k = d$, but that the <u>Young diagram</u> T <u>has more than one column</u>; without loss of generality, we may assume that both indices 1 and 2 occur in the first row. Here we consider $c_T(e_1 \otimes e_1 \otimes e_3 \otimes \dots \otimes e_d)$ and $c_T(e_2 \otimes e_2 \otimes e_3 \otimes \dots \otimes e_d) \in c_T(\otimes^d V)$, and find that the former has $e_1 \otimes e_1 \otimes e_3 \otimes \dots \otimes e_d$ present with coefficient 2, which is the coefficient of $e_2 \otimes e_2 \otimes e_3 \otimes \dots \otimes e_d$ in the latter, and our two elements of $c_T(\otimes^d V)$ are linearly independent. Therefore, if $1 \le k \le d$, <u>there are no abelian ideals in</u> $S^k(A)$, $A = \mathrm{End}_F(V)$, <u>except when</u> $k = d$ <u>and the ideal in question is</u> $\mathrm{End}_F((\sum_{\sigma \in S_k} \mathrm{sgn}(\sigma)\sigma)(\otimes^d V)) = \mathrm{End}_F(\Lambda^d V)$. <u>In this case, the abelian ideal is minimal and unique</u>.

When A is a general central simple algebra, $[A:F] = d^2$, it follows by field extension that $S^k(A)$ has no abelian ideals if $1 \le k \le d$. If B is an abelian ideal in $S^d(A)$, then B_K, for K a splitting field, is an abelian ideal in $S^d(A_K)$, so has dimension one and is unique by the above. That is, if $S^d(A)$ <u>has an abelian ideal, it is unique and has dimension one</u>.

To see that $S^d(A)$ actually contains an abelian ideal, we may take the splitting field K to be a finite Galois extension of F. If $\mathcal{B}$ is any minimal ideal in $S^d(A)$, and if I is a minimal ideal in $\mathcal{B}_K$, then so is each I^σ, $\sigma \in \mathrm{Gal}(K/F)$, and $\mathcal{B}_K = \sum_\sigma I^\sigma$, which is abelian if I is. Since $S^d(A_K) = \sum_{\mathcal{B}} \mathcal{B}_K$, some $\mathcal{B}_K$ must be abelian (and hence one-dimensional) by the split case, and therefore $\mathcal{B}$ is. We have shown:

PROPOSITION IV.3: Let A be central simple over F, $[A:F] = d^2$. Then $S^k(A)$ has no non-zero commutative ideals for $1 < k < d$, while $S^d(A)$ has a unique commutative ideal, which has dimension 1.

When A is involutorial of second kind, $[A:F] = 2d^2$, a similar argument applies; in the split case, no $S^r(Ae_1) \otimes S^s(Ae_2)$ has commutative ideals for $0 < r + s \le d$, unless $(r,s) = (d,0)$ or $(0,d)$. In each of these cases, there is a unique commutative ideal, of dimension 1. When A is not assumed to be split, with $Z = F(\zeta)$, $\zeta^2 = \gamma$ as at the beginning of this section, we see that if $\rho(\zeta)$ is the projection of $\rho_d(\zeta)$ on a commutative ideal $\mathcal{B}$ in $S^d(A)$ (the only $S^k(A)$, $1 \le k \le d$ where non-zero commutative ideals can be present), then $\rho(\zeta)^2 = d^2 \gamma 1_{\mathcal{B}}$. Moreover, $\mathcal{B}_K$ has dimension at most 2, and if $\xi \in Z$ is mapped into $\mathcal{B}$ by $\xi \to \rho(\xi)$ defined as was $\rho(\zeta)$ above, then the map $\varphi:\xi \to \frac{1}{d}\rho(\xi)$ sends 1 to the unit element of $\mathcal{B}$ and ζ to $\varphi(\zeta)$, where $\varphi(\zeta)^2 = \gamma 1_{\mathcal{B}}$. Thus φ defines a homomorphism of Z into $\mathcal{B}$, and by dimensions $\varphi(Z) = \mathcal{B}$. That is, if $S^k(A)$ has a commutative ideal for $1 \le k \le d$, then $k = d$, and there is a unique non-zero commutative ideal $\mathcal{B}$ in $S^k(A)$; $\mathcal{B}$ is isomorphic to the field Z by means of ρ, as above.

All the underlined assertions have been proved except for the existence of our non-zero commutative ideal. That is shown exactly as in the proof of Proposition 3, using a Galois extension which contains Z.

5. THE MINIMAL IDEALS OF $S^k(A)$ ARE CENTRAL SIMPLE.

The assumption is that A is central simple over F, $[A:F] = d^2$, and $k \le d$. The proof we give of the assertion of the title is based on an idea of Tamagawa, that has been developed further by Darrell Haile [9].

Let R_x be the transformation $y \to yx$ of A, for given $x \in A$, so the symmetric bilinear form

$$(u,v) = \mathrm{Tr}(R_u R_v)$$

is nondegenerate on A, and satisfies

$$(uw,v) = (u,wv) = (w,vu)$$

for all $u, v, w \in A$. Let $\{e_i\}$, $1 \le i \le d^2$, be a basis for A, and let $\{f_i\}$ be a basis with $(e_i,f_j) = d\delta_{ij}$ for all i,j. Then if $a \in A$,

$$e_i a = \sum_j \alpha_{ij} e_j, \quad af_k = \sum_\ell \beta_{k\ell} f_\ell,$$

then from $(e_i a, f_j) = (e_i, af_j)$ we have

(9) $$\alpha_{ij} = \beta_{ji}.$$

Likewise if $ae_i = \sum_j \gamma_{ij} e_j$, $f_k a = \sum_\ell \varepsilon_{k\ell} f_\ell$, then from $(ae_i, f_j) = (e_i, f_j a)$ it follows that

(10) $$\gamma_{ij} = \varepsilon_{ji}.$$

Now let $c = \sum_{i=1}^{d^2} e_i \otimes f_i \in A \otimes A$.

Then $$(a \otimes 1)c = \sum_i ae_i \otimes f_i = \sum_{i,j} \gamma_{ij} e_j \otimes f_i$$
$$= \sum_{i,j} \varepsilon_{ji} e_j \otimes f_i = \sum_{i,j} \varepsilon_{ij} e_i \otimes f_j = c(1 \otimes a),$$

and likewise $(1 \otimes a)c = c(a \otimes 1)$, for all $a \in A$. It follows that, for all $a, b \in A$,

(11) $$(a \otimes b)c = (a \otimes 1)(1 \otimes b)c = (a \otimes 1)c(b \otimes 1) = c(1 \otimes a)(b \otimes 1) = c(b \otimes a),$$

and that c^2 is central in $A \otimes A$. Now $A \otimes A$ is central simple, so $c^2 = \mu(1 \otimes 1)$, where $\mu \in F$.

In $A \otimes A$, we have $Tr(R_{c^2}) = \mu Tr(R_{1\otimes 1}) = \mu d^4$. On the other hand $Tr(R_{c^2})$

$$= Tr(R_{(\Sigma e_i \otimes f_i)c}) = Tr(R_{c\Sigma f_i \otimes e_i})$$
$$= Tr(R_{\sum_{i,j} e_j f_i \otimes f_j e_i}) = \sum_{i,j} Tr_A(R_{e_j} R_{f_i}) Tr_A(R_{f_j} R_{e_i})$$
$$= d^2 \sum_{i,j} \delta_{ij} = d^2 \sum_i 1 = d^4, \text{ and } \mu = 1: c^2 = 1 \otimes 1.$$

For $1 \le j < k$, let $s_j \in \otimes^k A$ be equal to

$$\overbrace{1 \otimes 1 \otimes \ldots \otimes 1}^{j-1} \otimes c \otimes \overbrace{1 \otimes \ldots \otimes 1}^{k-j-1}.$$ Then

$s_j^2 = 1 \in \otimes^k A$, and $(a_1 \otimes \ldots \otimes a_k)s_j = s_j(a_1 \otimes \ldots \otimes a_{j-1} \otimes a_{j+1} \otimes a_j \otimes a_{j+2} \otimes \ldots \otimes a_k)$. That is, each s_j is of multiplicative period two in $\otimes^k A$, and conjugation by s_j effects the transposition of the j-th and $(j+1)$-th tensor factors in $\otimes^k A$.

Clearly s_i and s_j commute except perhaps if $|i-j| = 1$. We show $s_i s_{i+1}$ has period 3; for this, it suffices to cube $(c \otimes 1)(1 \otimes c) \in \otimes^3 A$, and to show the result is 1. Now

$$((c \otimes 1)(1 \otimes c))^3 = (\sum_i e_i \otimes f_i \otimes 1)(1 \otimes c)((c \otimes 1)(1 \otimes c))^2$$
$$= (1 \otimes c)(\sum_i e_i \otimes 1 \otimes f_i)(c \otimes 1)(1 \otimes c)(c \otimes 1)(1 \otimes c)$$
$$= (1 \otimes c)(c \otimes 1)(\sum_i 1 \otimes e_i \otimes f_i)(1 \otimes c)(c \otimes 1)(1 \otimes c)$$
$$= (1 \otimes c)(c \otimes 1)(1 \otimes c)^2(c \otimes 1)(1 \otimes c)$$
$$= (1 \otimes c)(c \otimes 1)^2(1 \otimes c) = (1 \otimes c)^2 = 1 \otimes 1 \otimes 1,$$

and clearly $(c \otimes 1)(1 \otimes c) \neq 1 \otimes 1 \otimes 1$, since $d \geq 3$.

Now the symmetric group S_k is defined by generators w_i, $1 \leq i \leq k-1$, the transpositions of i and $i + 1$, with relations: $w_i^2 = 1$; $w_i w_j = w_j w_i$ if $|i-j| > 1$; $(w_i w_{i+1})^3 = 1$. By considering the homomorphism

$$S_k \to \langle s_1, \ldots, s_{k-1} \rangle = G \text{ sending } w_i \text{ to } s_i,$$

whose existence is assured by the relations on the s_i, and combining this with the map $G \to \mathrm{Aut}\,(\otimes^k A)$ sending s to its inner automorphism of $\otimes^k A$, <u>we see that G is a group of units in $\otimes^k A$, isomorphic to S_k</u>, whose centralizer in $\otimes^k A$ is $S^k(A)$.

In particular, the subalgebra FG of $S^k(A)$ consisting of the linear combinations of elements of G is a homomorphic image of the group algebra $F[S_k]$, and has centralizer $S^k(A)$ in $\otimes^k A$. It follows from the semi-simplicity of $F[S_k]$ that FG <u>is a semisimple algebra</u>.

Upon passage to a splitting field K for A, where $A_K = \mathrm{End}_K(V)$, $[V:K] = d$, it is still true (by linear algebra) that the centralizer in $\otimes^k A_K = \mathrm{End}_K(\otimes^k V)$ of $KG = (FG)_K$ is $S^k(A)_K = S^k(A_K)$, and KG is still semisimple. That is,

$$S^k(A)_K = \mathrm{Hom}_{KG}(\otimes^k V, \otimes^k V)$$

Now KG is semisimple and $\otimes^k V$ is a faithful, finitely generated KG-module, so has the double centralizer property (cf. [7], p. 405):

$$\mathrm{Hom}_{S^k(A)_K}(\otimes^k V, \otimes^k V) = KG.$$

From §1 we know that this ring of endomorphisms must also be $K[S_k]$; thus $KG = K[S_k]$, <u>and the elements of G are linearly independent over K, therefore also over F. That is, $FG \cong F[S_k]$, a semisimple F-algebra whose center we have determined in §1: The center C of $F[S_k]$ is the direct sum of $p(k)$ fields isomorphic to F.</u>

Clearly $C \subseteq S^k(A)$ and is central in $S^k(A)$. Thus the center of $S^k(A)$ contains C, with $[C:F] = p(k)$. Again from §1, $S^k(A)$ has center of dimension exactly $p(k)$, so that C is the center of $S^k(A)$. Now the center of $S^k(A)$ is the sum of $p(k)$ fields isomorphic to F, and this <u>means</u>

that $S^k(A)$ is the sum of $p(k)$ minimal ideals, each of them a central simple algebra over F.

Still with $k \le d$, my colleague David Saltman has supplied me with more information about these central simple minimal ideals: The central simple algebra $\otimes^k A$ has the form $M_t(\mathcal{D})$, where $\mathcal{D}$ is a central division algebra, uniquely determined. Then each of the minimal ideals in $S^k(A)$ has the form $M_j(\mathcal{D})$ for suitable j and this same division algebra $\mathcal{D}$. In particular, when $k = d$ we have $\mathcal{D} = F$, since the exponent of $[A]$ in the Brauer group divides d. In this case each minimal ideal in $S^d(A)$ is a split algebra, and all irreducible $S^k(A)$-modules are absolutely irreducible. When $\mathcal{D}$ is involutorial (of first kind) over F, $A = M_r(\mathcal{D})$, then $[\otimes^k A] = 1$ or $[\mathcal{D}]$, according as k is even or odd; the implications for $S^k(A)$ are clear. Saltman has also developed the ideas of this section within the more general context of Azumaya algebras. His work is not yet published.

CHAPTER V

CONSTRUCTION OF REPRESENTATIONS: TYPE A AND TYPE C (FIRST KIND)

If $r > 2$, a central simple Lie algebra L of relative type A_r over F is isomorphic to a Lie algebra $\mathfrak{sl}(r+1, \mathcal{D})$, the derived algebra of the Lie algebra of $(r+1)$ by $(r+1)$ matrices over the central (associative) division algebra $\mathcal{D}$. L is split if and only if $\mathcal{D} = F$; thus we assume $[\mathcal{D}:F] = d^2 \geq 4$.

If $r > 3$, a central simple Lie algebra L of relative type C_r over F is either: 1) isomorphic to the Lie algebra of skew transformations in a vector space V of dimension $2r$ over $\mathcal{D}$, with respect to an involution as follows: $\mathcal{D}$ has an involution over F (thus "of first kind"), and V carries a nondegenerate antihermitian form of Witt index r; the involution in $\mathrm{End}_{\mathcal{D}}(V)$ is the transpose with respect to the form. Or: 2) L is associated with an involution of second kind, a case that will be dealt with in the next chapter.

The cases described in detail above have many common features. We bring them together here for that reason, including all values of $r \geq 1$. The principles of the earlier chapters are invoked to construct all their finite-dimensional F-irreducible representations, as parametrized by relative highest weights.

1. THE LIE ALGEBRAS AND THEIR FUNDAMENTAL WEIGHTS.

When $\mathcal{D}$ is a central division algebra over F, $[\mathcal{D}:F] = d^2 \geq 4, \mathfrak{sl}(n,\mathcal{D})$ is used to denote the derived algebra of the Lie algebra $M_n(\mathcal{D})$ of n by n $\mathcal{D}$-matrices, i.e., the Lie algebra of all n by n $\mathcal{D}$-matrices (a_{ij}) such that $\sum_{i=1}^{n} a_{ii} \in [\mathcal{D}\mathcal{D}]$. We denote by T the set of diagonal matrices in $\mathfrak{sl}(n,\mathcal{D})$ with entries in F, i.e., the set of n by n F-diagonal matrices of trace zero, a maximal split torus in $\mathfrak{sl}(n,\mathcal{D})$, with $T \neq 0$ if and only if $n > 1$. The root-spaces relative to T are the spaces $\mathcal{D} E_{ij}$, $i \neq j$, and we may take as positive roots those for which the corresponding space has $i < j$; then the roots $\alpha_1, \ldots, \alpha_{n-1}$ to which $\mathcal{D}E_{12}, \mathcal{D}E_{23}, \ldots, \mathcal{D}E_{n-1,n}$ belong are the simple roots. The roots for $\mathcal{D}E_{ij}$ and $\mathcal{D}E_{ji}$ are negatives of one another. The centralizer L_0 of T is the set of all diagonal ($\mathcal{D}$-)matrices in $\mathfrak{sl}(n,\mathcal{D})$.

In Appendix 1, we display a splitting of $\mathfrak{sl}(n,\mathcal{D})$ having the properties of Chapter II. Over the splitting field K, the split Lie algebra is $\mathfrak{sl}(nd,K)$, and a fundamental system of roots $\gamma_1, \ldots, \gamma_{nd-1}$ as in Chapter II has $\gamma_i|T = 0$ if $d \nmid i$, while $\gamma_{dj}|T = \alpha_j$, $1 \le j \le n-1$. The fundamental weights for $\mathfrak{sl}(nd, K)$ associated with γ_i, say $\pi_1, \ldots, \pi_{nd-1}$, have $\pi_{dj+k}|T = (d-k)\lambda_j + k\lambda_{j+1}$, $0 \le j \le n-1$, $0 \le k \le d$, where $\lambda_1, \ldots, \lambda_{n-1}$ are the fundamental dominant integral functions on T associated with $\alpha_1, \ldots, \alpha_{n-1}$ and where $\lambda_0 = \lambda_n = 0$. Thus the highest weight λ of any irreducible $\mathfrak{sl}(n,\mathcal{D})$-module is a non-negative integral combination of these restrictions. Furthermore, Proposition II.3 shows that the irreducible modules of highest weight λ are those obtained by Cartan multiplication from those whose highest weight is a restriction of some π_i.

We organize these restrictions into three groups:

A): $d\lambda_j$, $1 < j < n-1$;

B): $k\lambda_j$, $1 \le k \le d$, $j = 1$ or $n-1$;

C): $(d-k)\lambda_j + k\lambda_{j+1}$, $1 \le j < n-1$, $1 \le k < d$.

Next suppose the central division algebra $\mathcal{D}$ has an involution $a \mapsto a^*$ over F. Let an antihermitian form $(\cdot,\cdot)$ on a vector space V of dimension 2n over $\mathcal{D}$ be as in the introduction. Then V has a $\mathcal{D}$-basis $e_1, \ldots, e_{2n}$, with $(e_i,e_j) = \varepsilon_{ij}\,\delta_{i,2n+1-j}$, where $\varepsilon_{ij} = 1$ if $i \le j$, $\varepsilon_{ij} = -1$ if $i > j$. To fix our notion of antihermitian form, we assume $(u,v) = -(v,u)^*$; $(au,v) = a(u,v)$ for all $u, v \in V$, $a \in \mathcal{D}$. Then $(u,av) = (u,v)a^*$. Assuming $n \ge 1$, we let L be the set of all $\mathcal{D}$-endomorphisms T of V such that $(uT,v) + (u,vT) = 0$ for all $u, v \in V$. Then L is a Lie algebra over F, and is central simple except when $n = 1$ and $\mathcal{D}$ is a quaternion algebra with the usual involution.

Relative to the basis $e_1, \ldots, e_{2n}$ for V over $\mathcal{D}$, we may describe L as the set of 2n by 2n $\mathcal{D}$-matrices (a_{ij}) satisfying the following conditions: $a_{ij}^* = -a_{2n+1-j,2n+1-i}$; $a_{i,2n+1-j}^* = a_{j,2n+1-i}$; $a_{2n+1-i,j}^* = a_{2n+1-j,i}$ for $1 \le i, j \le n$. In particular, $a_{i,2n+1-i}$ is fixed under the involution for $1 \le i \le 2n$. In this identification, a maximal split torus T in L consists of all F-diagonal matrices in L, and is spanned by all $E_{ii} - E_{2n+1-i,2n+1-i}$, $1 \le i \le n$. The centralizer L_0 of T consists of all diagonal $\mathcal{D}$-matrices in L, i.e., all matrices

$$\sum_{i=1}^{n} (a_i E_{ii} - a_i^* E_{2n+1-i,2n+1-i}),$$

$a_i \in \mathcal{D}$.

The root-spaces L_α of L relative to T are of three kinds:

i) $\{aE_{ij} - a^*E_{2n+1-j,2n+1-i} \mid a \in \mathcal{D}\}$, $1 \le i, j \le n$, $i \ne j$;

ii) $\{aE_{i,2n+1-j} + a^*E_{j,2n+1-i} \mid a \in \mathcal{D}\}$, $1 \le i \le j \le n$;

iii) $\{aE_{2n+1-i,j} + a^*E_{2n+1-j,i} \mid a \in \mathcal{D}\}$, $1 \le i \le j \le n$;

(When $i = j$ in ii) and iii), these are respectively those matrices

$$aE_{i,2n+1-i},\ aE_{2n+1-i,i} \text{ with } a^* = a.$$

A set of positive roots relative to $\mathcal{T}$ consists of those $\alpha \in \mathcal{T}^*$ to which the root-vectors of ii) belong, as well as those of i) with $i < j$; their negatives have as root-vectors the ones listed under iii) and those of i) with $i > j$, respectively. With an ordering of this kind, the simple roots are as follows:

α_i, $1 \le i < n$, with root-space $\{aE_{i,i+1} - a^*E_{2n-i,2n+1-i} \mid a \in \mathcal{D}\}$;

α_n, with root-space $\{aE_{n,n+1} \mid a \in \mathcal{D},\ a^* = a\}$.

The splitting of L over an extension field K results in a split simple Lie algebra of type C_{nd} or D_{nd}, according as the involution $a \mapsto a^*$ in $\mathcal{D}$ is of "orthogonal" or "symplectic" type, i.e., according as the set of fixed elements has dimension $\binom{d+1}{2}$ or $\binom{d}{2}$. A fundamental system of roots $\gamma_1, \ldots, \gamma_{nd}$ will be displayed in Appendix 2, with $\gamma_i|\mathcal{T} = 0$ if $d \nmid i$, while $\gamma_{dj}|\mathcal{T} = \alpha_j$, as above. Then the fundamental weights $\pi_1, \ldots, \pi_{nd}$ for L_K associated with $\gamma_1, \ldots, \gamma_{nd}$ have $\pi_{dj+k}|\mathcal{T} = (d-k)\lambda_j + k\lambda_{j+1}$ as for $\mathfrak{sl}(n,\mathcal{D})$, when $0 \le j < n-1$, $0 \le k \le d$; also $\pi_{(n-1)d+k}|\mathcal{T} = (d-k)\lambda_{n-1} + k\lambda_n$, $0 \le k < d-2$. Finally, for $\pi_{nd-2}, \pi_{nd-1}, \pi_{nd}$ a distinction must be drawn between the cases where the involution $*$ in $\mathcal{D}$ is of orthogonal and symplectic type: When $*$ is orthogonal, we have

$$\pi_{(n-1)d+k}|\mathcal{T} = (d-k)\lambda_{n-1} + k\lambda_n, \quad 0 \le k \le d;$$

when $*$ is symplectic,

$$\pi_{nd-2}|\mathcal{T} = 2\lambda_{n-1} + (d-2)\lambda_n;\ \pi_{nd-1}|\mathcal{T} = \lambda_{n-1} + (\tfrac{d}{2} - 1)\lambda_n;$$

$$\pi_{nd}|\mathcal{T} = \frac{d}{2}\lambda_n.$$

Again by Proposition II.3 all irreducible modules result from those for these (relative) highest weights by Cartan multiplication. The restrictions we have determined are again organized below into groups:

A) $d\lambda_j$, $1 < j \le n-1$;

B) $k\lambda_1$, $1 \le k \le d$ ($k \le d-2$ if $*$ is symplectic and $n = 1$);

C) $(d-k)\lambda_j + k\lambda_{j+1}$, $1 \le j < n-1$, $1 \le k < d$;

D_1) ($*$ orthogonal) $d\lambda_n$;

D_2) (* symplectic) $\frac{d}{2}\lambda_n$;
E_1) (* orthogonal) $(d-k)\lambda_{n-1} + k\lambda_n$, $1 \le k < d$;
E_2) (* symplectic) $(d-k)\lambda_{n-1} + k\lambda_n$, $1 \le k < d-1$;
F) (* symplectic) $\lambda_{n-1} + (\frac{d}{2} -1)\lambda_n$.

We shall also need the elements $H_i \in T$ with $H_i = [xy]$, $x \in L_{\alpha_i}$, $y \in L_{-\alpha_i}$, $\alpha_i(H_i) = 2$. For $s\ell(n,\mathcal{D})$,

$$H_i = E_{i+1,i+1} - E_{ii}, \quad 1 \le i \le n-1.$$

For the Lie algebras L of relative type C_n,

$$H_i = E_{i+1,i+1} - E_{i,i} - E_{2n-i,2n-i} + E_{2n+1-i,2n+1-i},$$

$1 \le i \le n-1$, while

$$H_n = E_{n+1,n+1} - E_n.$$

(All the information on $s\ell(n,\mathcal{D})$ and on L may be easily verified; what relates to splitting is explained in the Appendices.)

2. CONSTRUCTION OF REPRESENTATIONS: WEIGHTS OF GROUP A.

For all these weights λ, we have $\lambda(H_1) = 0$. In terms of Chapter I, any λ-admissible irreducible L_0-module must be annihilated by $\psi(e_{-\alpha_1} e_{\alpha_1}) \in U(L_0)$ for each $e_{-\alpha_1} \in L_{-\alpha_1}$, $e_{\alpha_1} \in L_{\alpha_1}$. Since $e_{-\alpha_1}e_{\alpha_1} = [e_{-\alpha_1}, e_{\alpha_1}] + e_{\alpha_1} e_{-\alpha_1}$, we have $\psi(e_{-\alpha_1} e_{\alpha_1}) = [e_{-\alpha_1}, e_{\alpha_1}] \in L_0$. Now for $s\ell(n,\mathcal{D})$, the typical element of $L_{-\alpha_1}$ is aE_{21}, $a \in \mathcal{D}$, and the typical element of L_{α_1} is bE_{12}, $b \in \mathcal{D}$. Thus the condition of annihilation of the irreducible L_0-module W by all $[e_{-\alpha_1} e_{\alpha_1}]$ amounts to saying that W is annihilated by all $abE_{22} - baE_{11}$, $a, b \in \mathcal{D}$. With $b = 1$, this says that W is annihilated by all $aE_{22} - aE_{11}$, $a \in \mathcal{D}$, and therefore, upon forming brackets, by all $[ab]E_{22} + [ab]E_{11}$, $a, b \in \mathcal{D}$. Combining this with the annihilation of W by all $[ab]E_{22} - [ab]E_{11}$, we see that W is annihilated by all cE_{22} and by all cE_{11}, for $c \in [\mathcal{DD}]$. Now every element a of $\mathcal{D}$ can be uniquely written $a = \tau(a)1 + a_0$, $\tau(a) \in F$ and $a_0 \in [\mathcal{DD}]$. Thus the action on W of

$$\sum_{i=1}^{n} a_i E_{ii} \in L_0, \quad \sum_i a_i \in [\mathcal{DD}], \quad \underline{i.e.}, \quad \sum_i \tau(a_i) = 0,$$

is the same as that of

$$\tau(a_1)E_{11} + \tau(a_2)E_{22} + \sum_{i=3}^{n} a_iE_{ii}, \quad \text{or of}$$
$$\tau(a_1+a_2)E_{22} + \sum_{i=3}^{n} a_iE_{ii} .$$

In the case of the Lie algebra L of skew transformations, the typical element of $L_{-\alpha_1}$ is $aE_{21} - a^*E_{2n,2n-1}$, while that of L_{α_1} is $bE_{12} - b^*E_{2n-1,2n}$. Their bracket, which must annihilate a λ-admissible L_0-module W, is

$$abE_{22} - baE_{11} + (ba)^*E_{2n,2n} - (ab)^*E_{2n-1,2n-1}.$$

Again W must be annihilated by all $cE_{22} - c^*E_{2n-1,2n-1}$ and by all $cE_{11} - c^*E_{2n,2n}$, $c \in [\mathcal{D}\mathcal{D}]$, and the action on W of the general element $\sum_{i=1}^{n} -a_i^* E_{ii} + a_iE_{2n+1-i,2n+1-i}$ of L_0 $(a_i \in \mathcal{D})$ is the same as that of

$$-\tau(a_1+a_2)E_{22} + \tau(a_1+a_2)E_{2n-1,2n-1} + \sum_{i=3}^{n} (-a_i^* E_{ii} + a_iE_{2n+1-i,2n+1-i}).$$

Repetition of this argument yields the following:

LEMMA V.1. <u>Let</u> $j \geq 1$, <u>and let</u> W <u>be a</u> λ-<u>admissible</u> L_0-<u>module, where</u> $\lambda(H_i) = 0$ <u>for all</u> $i \leq j$. <u>Then, if</u> $L = s\ell(n,\mathcal{D})$, <u>the action on</u> W <u>of the general element</u>

$$\sum_{i=1}^{n} a_iE_{ii} \quad \text{of} \quad L_0, \quad \sum_i \tau(a_i) = 0,$$

<u>agrees with that of</u>

$$\tau(a_1 +\ldots+ a_{j+1})E_{j+1,j+1} + \sum_{i=j+2}^{n} a_iE_{ii} .$$

<u>In the other case under consideration, the action on</u> W <u>of the general element</u> $\sum_{i=1}^{n} - a_i^*E_{ii} + a_iE_{2n+1-i,2n+1-i}$ <u>of</u> L_0 <u>agrees with that of</u>

$$\tau(a_1 +\ldots+ a_{j+1})(E_{2n-j,2n-j} - E_{j+1,j+1})$$
$$+ \sum_{i=j+2}^{n} - a_i^*E_{ii} + a_iE_{2n+1-i,2n+1-i} .$$

For $s\ell(n,\mathcal{D})$, the same argument shows that if $\lambda(H_i) = 0$ for $i = j + 1, j + 2,\ldots, n-1$, then the action on W of Σa_iE_{ii} agrees with that of $\sum_{i=1}^{j-1} a_iE_{ii} + \tau(a_j +\ldots+ a_n)E_{jj}$. In the other case, let W be λ-admissible for L_0 with $\lambda(H_n) = 0$. Then for all $a, b \in \mathcal{D}$, $a^* = a, b^* = b$, W must be annihilated by $[aE_{n+1,n}, bE_{n,n+1}]$

$$= abE_{n+1,n+1} - baE_{nn}.$$

Again for $b = 1$, this gives annihilation by all $aE_{n+1,n+1} - aE_{nn}$ with $a^* = a$; forming commutators gives annihilation by all $[ab]E_{n+1,n+1} + [ab]E_{nn} = [ab]E_{n+1,n+1} - [ab]^*E_{n,n}$ for $a^* = a, b^* = b$. Now the elements $[ab]$ for $a^* = a, b^* = b$, span the *-skew elements of $\mathcal{D}$ unless $d = 2$ and the *-fixed elements are the center of $\mathcal{D}$, i.e., unless $\mathcal{D}$ is a quaternion algebra with standard involution. That is, _except when $\mathcal{D}$ is a quaternion algebra with standard (symplectic) involution, $\mathcal{W}$ is annihilated by all_ $-a^*E_{nn} + aE_{n+1,n+1}$, $a \in \mathcal{D}$. With this exception, it now follows as before that if $\lambda(H_i) = 0$ for $i = j+1, \ldots, n$ then $\mathcal{W}$ is annihilated by all $-a^*E_{ii} + aE_{2n+1-i,2n+1-i}$, $a \in \mathcal{D}$, $j+1 \le i \le n$.

In the exceptional case, suppose also that $\lambda(H_{n-1}) = 0$. Then $\mathcal{W}$ is annihilated by all $[-a^*E_{n,n-1} + aE_{n+2,n+1}, -b^*E_{n-1,n} + bE_{n+1,n+2}]$ $= a^*b^*E_{nn} - b^*a^*E_{n-1,n-1} + abE_{n+2,n+2} - baE_{n+1,n+1}$, $a, b \in \mathcal{D}$. By the argument that led to Lemma 1, $\mathcal{W}$ is annihilated by all $-c^*E_{nn} + cE_{n+1,n+1}$ and by all $-c^*E_{n-1,n-1} + cE_{n+2,n+2}$, for $c \in [\mathcal{D}\mathcal{D}]$. Furthermore, the action of $-E_{n-1,n-1} + E_{n+2,n+2}$ agrees with that of $-E_{nn} + E_{n+1,n+1}$, which we have seen to be zero. Thus $\mathcal{W}$ is annihilated in this case by all $-a^*E_{nn} + aE_{n+1,n+1}$ and by all $-a^*E_{n-1,n-1} + aE_{n+2,n+2}$, $a \in \mathcal{D}$. These observations are summarized and extended in an obvious way as follows:

LEMMA V.2. Let $\mathcal{W}$ be a λ-admissible L_0-module where $\lambda(H_i) = 0$ for all $i \ge j$. If $L = s\ell(n,\mathcal{D})$, let $j \le n-1$. Then the action on $\mathcal{W}$ of the general element $\Sigma_i a_i E_{ii}$, $\Sigma\tau(a_i) = 0$, of L_0 agrees with that of

$$\sum_{i=1}^{j-1} a_i E_{ii} + \tau(a_j + \ldots + a_n)E_{jj} .$$

In the other case, let $j \le n$, except that $j \le n-1$ if $\mathcal{D}$ is quaternionic with standard involution. Then the action on $\mathcal{W}$ of

$$\sum_{i=1}^{n} (-a_i^*E_{ii} + a_iE_{2n+1-i,2n+1-i}) \in L_0 \quad \text{agrees with that of}$$

$$\sum_{i=1}^{j-1} (-a_i^*E_{ii} + a_iE_{2n+1-i,2n+1-i}) .$$

Combining Lemmas 1 and 2, we see that if $\mathcal{W}$ is a $k\lambda_j$-admissible L_0-module, with $1 < j < n$ (with $1 < j < n-1$ if $\mathcal{D}$ is quaternionic with standard involution) then the action on $\mathcal{W}$ of the general element of L_0,

(1) $\sum_{i=1}^{n} a_iE_{ii}$, $\Sigma\tau(a_i) = 0$, for $s\ell(n,\mathcal{D})$;

(2) $\sum_{i=1}^{n} (-a_i^* E_{ii} + a_i E_{2n+1-i,2n+1-i})$, otherwise, is the same, respectively as that of

$$\tau(a_1 + \ldots + a_j)E_{jj} + \tau(a_{j+1} + \ldots + a_n)E_{j+1,j+1}$$

$$= \tau(a_{j+1} + \ldots + a_n)H_j, \quad \text{or of}$$

$$\tau(a_1 + \ldots + a_j)(E_{2n+1-j,2n+1-j} - E_{jj}).$$

In the case of $\mathfrak{sl}(n,\mathcal{D})$, H_j acts in $\mathcal{W}$ by scalar multiplication by $k\lambda_j(H_j) = k$. In the latter, since $E_{2n-j,2n-j} - E_{j+1,j+1}$ annihilates $\mathcal{W}$, the action of $E_{2n+1-j,2n+1-j} - E_{jj}$ is the same as that of H_j, and again is multiplication by k. Thus we have

LEMMA V.3. Let $\mathcal{W}$ be a $k\lambda_j$-admissible L_0-module with $1 < j < n$ (with $1 < j < n-1$ in the case of type C_n, where $\mathcal{D}$ is quaternionic with standard involution). Then the action on $\mathcal{W}$ of the general element (1) of $L_0 \subseteq \mathfrak{sl}(n,\mathcal{D})$ is scalar multiplication by $k\tau(a_{j+1} + \ldots + a_n)$; and otherwise that of the general element (2) of L_0 is scalar multiplication by $k\tau(a_1 + \ldots + a_j)$. Thus if $\mathcal{W}$ is irreducible, $\mathcal{W}$ is one-dimensional, therefore absolutely irreducible, and any irreducible L-module of highest weight $k\lambda_j$ is unique and absolutely irreducible.

The last assertions follow from the considerations of Chapters I and II. It is perhaps worth remarking as to the implications of our considerations when $\lambda = 0$: For $n \geq 4$ in general we may think of λ as $0 \cdot \lambda_2$, and conclude that $\mathcal{W}$ must be one-dimensional, annihilated by L_0, and again *there is a unique irreducible L-module of highest weight 0; this is obviously the trivial L-module* F.

In fact, one can relax the restrictions on n for $\lambda = 0$: Evidently the same argument applies when $n = 3$ in the involutory case, $\mathcal{D}$ not being the quaternionic exception. For $n = 2,3$, Lemma 1 for $\mathfrak{sl}(n,\mathcal{D})$ shows that the action of the element (1) on a 0-admissible L_0-module $\mathcal{W}$ is the same as that of, respectively,

$$\tau(a_1 + a_2)E_{22} = 0, \quad \tau(a_1 + a_2 + a_3)E_{33} = 0,$$

and again $\mathcal{W}$ is a trivial L_0-module. Likewise, in the involutory case, Lemma 1 shows that the action of (2) is the same as that of $\tau(a_1 + \ldots + a_n)(E_{n+1,n+1} - E_{nn}) = \tau(a_1 + \ldots + a_n)H_n$, which is zero since $\lambda(H_n) = 0$. This argument applies for $n \geq 2$. When $n = 1$, we have seen in §1 that the exceptional quaternionic case is *not* central simple; excluding this, Lemma 2 shows that the general element of L_0 annihilates $\mathcal{W}$. Thus *the underlined assertion of the last paragraph holds for all central simple Lie algebras* L *under consideration in this chapter.*

A note on the exceptional case for $n = 1$ may be in order: Here L is not even simple over F, being the sum of the ideal L_1 of 2 by 2 F-matrices of trace zero and the ideal L of all $aE_{11} + aE_{22}$, $a^* = -a$ (*i.e.*, a is a "pure vector" quaternion). To say W is 0-admissible is simply to say that W is annihilated by L_1 ; any irreducible action of L_2 can be combined with the trivial action of L_1 to yield a module we should have to acknowledge as "0-admissible".

With $\lambda = k\lambda_j$ as in Lemma 3, $k > 0$, it follows from Lemma 3 that every element of $U(L_0)$ is congruent to a scalar, modulo the ideal $K(\lambda)$ of §1.4, so that $\dim U(L_0)/K(\lambda) \leq 1$. We next show that $K(\lambda) \neq U(L_0)$, so that there actually is an irreducible representation of highest weight $k\lambda_j$, if and only if $d|k$. For this we consider the condition that W be annihilated by all $\psi(e_{-\alpha_j}^{k+1} e_{\alpha_j}^{k+1}) \in U(L_0)$, where $e_{\pm\alpha_j} \in L_{\pm\alpha_j}$. Here, in $s\ell(n,\mathcal{D})$,

$e_{-\alpha_j} = aE_{j+1,j}$, $e_{\alpha_j} = bE_{j,j+1}$ for $a, b \in \mathcal{D}$, while otherwise

$e_{-\alpha_j} = -a^*E_{j+1,j} + aE_{2n+1-j,2n-j}$, $e_{\alpha_j} = -b^*E_{j,j+1} + bE_{2n-j,2n-j+1}$.

We consider first the case $b = 1$, and we generalize the first factor somewhat; namely, for commuting $a_1,\ldots, a_{k+1} \in \mathcal{D}$, we consider the effect on W of

(3) $\psi((a_1E_{j+1,j})(a_2E_{j+1,j})\ldots(a_{k+1}E_{j+1,j})(E_{j,j+1})^{k+1})$

resp.

(4) $\psi((-a_1^*E_{j+1,j} + a_1E_{2n+1-j,2n-j})\ldots(-a_{k+1}^* E_{j+1,j} + a_{k+1}E_{2n+1-j,2n-j})$

$(-E_{j,j+1} + E_{2n-j,2n+1-j})^{k+1})$. In either case, we denote by $f_{k+1}(a_1,\ldots, a_{k+1})$ the linear transformation of our proposed W which is the action of (3) or (4), even for non-commuting $a_1,\ldots, a_{k+1}$; we further define $f_t(a_1,\ldots, a_t)$ for each $t \geq 0$ as the action on W of the counterpart of (3) resp. (4) with $k + 1$ replaced by t (W still, however, being assumed $k\lambda_j$-admissible and irreducible).

Now $f_1(a)$ is the action on W of

$[aE_{j+1,j}, E_{j,j+1}]$ resp.

$[-a^*E_{j+1,j} + aE_{2n+1-j,2n-j}, -E_{j,j+1} + E_{2n-j,2n+1-j}]$,

i.e., of $aE_{j+1,j+1} - aE_{jj}$ resp.

$a^*E_{j+1,j+1} - a^*E_{jj} + aE_{2n+1-j,2n+1-j} - aE_{2n-j,2n-j}$,

thus of $k\tau(a)$ in either case: $f_1(a) = k\tau(a)$.

Now let $\rho(a) = k\tau(a) \in \mathrm{End}_F(\mathcal{W}) = F$, where $\mathcal{W}$ is an irreducible $k\lambda_j$-admissible L_0-module, necessarily of dimension 1. Then $\rho(1) = k1$, $\rho([ab]) = 0 = [\rho(a), \rho(b)]$ for all $a,b \in \mathcal{D}$, and we have seen in §III.4 that ρ satisfies the $k+1$-th symmetric identity if and only if $d|k$.

On the other hand, we shall show below that the sequence $f_1, f_2, \ldots$ satisfies our fundamental recursion of (1) of Chapter III when commuting arguments are substituted, and where $\rho(a) = k\tau(a)$. The explicit formula (6) of Chapter III for $f_{k+1}(a,a,\ldots, a)$ and the fact that f_{k+1} must vanish identically then shows that ρ must necessarily satisfy the (k+1)-th symmetric identity, hence that <u>there is no irreducible</u> $k\lambda_j$-<u>admissible</u> L_0-<u>module if</u> $d \nmid k$. (This may also be seen from the split viewpoint, by examining the possible restrictions to T of the dominant integral weights for L_K.)

For the first assertion of the last paragraph, let $t \geq 1$, and let $a_1,\ldots, a_{t+1} \in \mathcal{D}$ commute. In $\mathcal{U}(L)$,

$$
\begin{aligned}
(5)\qquad & (a_1E_{j+1,j})\cdots(a_{t+1}E_{j+1,j})(E_{j,j+1})^{t+1} \\
&= (E_{j,j+1})(a_1E_{j+1,j})\cdots(a_{t+1}E_{j+1,j})(E_{j,j+1})^{t} \\
&+ \sum_{i=1}^{t+1} (a_1E_{j+1,j})\cdots(a_iE_{j+1,j+1}-a_iE_{jj})\cdots(a_{t+1}E_{j+1,j})(E_{j,j+1})^{t} \\
&= (E_{j,j+1})(a_1E_{j+1,j})\cdots(a_{t+1}E_{j+1,j})(E_{j,j+1})^{t} \\
&+ \sum_{i=1}^{t+1} (a_iE_{j+1,j+1}-a_iE_{jj})(a_1E_{j+1,j})\cdots\widehat{(a_iE_{j+1,j})}\cdots(a_{t+1}E_{j+1,j})(E_{j,j+1})^{t} \\
&- \sum_{1\leq \ell < i \leq t+1} 2(a_\ell a_iE_{j+1,j})(a_1E_{j+1,j})\cdots\widehat{(a_\ell E_{j+1,j})}\cdots\widehat{(a_iE_{j+1,j})}\cdots \\
&\qquad \cdots (a_{t+1}E_{j+1,j})(E_{j,j+1})^{t}
\end{aligned}
$$

using the commutativity of $aE_{j+1,j}$ and $bE_{j+1,j}$, as well as that of a_ℓ and a_i. Applying ψ to (5) we see that the first term vanishes, and the action on $\mathcal{W}$ of the remaining terms is equal to

$$
\sum_{i=1}^{t+1} \rho(a_i)f_t(a_1,\ldots, \hat{a}_i,\ldots, a_{t+1})
$$
$$
- 2 \sum_{\ell < i} f_t(a_\ell a_i,a_1,\ldots, \hat{a}_\ell,\ldots, \hat{a}_i,\ldots, a_{t+1}).
$$

From our definition of f_{t+1}, the left side of (5) is $f_{t+1}(a_1,\ldots, a_{t+1})$. Thus we have seen that, in the case $L = s\ell(n,\mathcal{D})$, our sequence $f_1,f_2,\ldots$ associated with ρ satisfies the fundamental recursion (1) of Chapter III.

In the involutory case, exactly the same argument applies, with the same conclusion.

Conversely, as we have seen in III.4 we do have, for $\rho(a) = k\tau(a)$ with $d|k$, the (k+1)-th symmetric identity. For our purposes it is $k = d$ that is of interest here. We want to show that a one-dimensional space $\mathcal{W}$, on which the general element (1) or (2) of L_0 acts by $d\tau(a_{j+1} + \ldots + a_n)$ resp. $d\tau(a_1 + \ldots + a_j)$, is a $d\lambda_j$-admissible module in the sense of Chapter I. For this one must have, in $s\ell(n,\mathcal{D})$, that $[aE_{i+1,i}, bE_{i,i+1}]$ annihilates $\mathcal{W}$ for all $a,b \in \mathcal{D}$, $i \neq j$, and that $\psi((aE_{j+1,j})^{d+1} (bE_{j,j+1})^{d+1})$ annihilates $\mathcal{W}$. For the involutorial case, $\mathcal{W}$ must be annihilated by all

$$[-a^*E_{i+1,i} + aE_{2n+1-i,2n-i}, -b^*E_{i,i+1} + bE_{2n-i,2n+1-i}]$$

for all $i \neq j$, by all $[aE_{n+1,n}, bE_{n,n+1}]$ for $a^* = a, b^* = b$, and by all

$$\psi((-a^*E_{j+1,j} + aE_{2n+1-j,2n-j})^{d+1}(-b^*E_{j,j+1} + bE_{2n-j,2n+1-j})^{d+1})$$

for $a,b \in \mathcal{D}$. One can also check easily that $\mathcal{W}$ is an L_0-module on which $H \in T$ acts by $(d\lambda_j)(H)$.

For the admissibility conditions with $i \neq j$, we have, for instance if $i < j$,

$$\begin{aligned}&[-a^*E_{i+1,i} + aE_{2n+1-i,2n-i}, -b^*E_{i,i+1} + bE_{2n-i,2n+1-i}]\\ &= (ba)^*E_{i+1,i+1} - (ba)E_{2n-i,2n-i} - (ab)^*E_{ii} + (ab)E_{2n+1-i,2n+1-i},\end{aligned}$$

which acts on $\mathcal{W}$ by $d\tau(ab-ba) = 0$. All the other verifications not involving j are as easy as this, or easier. For the remaining ones, we show that if $1 \leq t \leq d + 1$, <u>and if</u> $a_1,\ldots, a_t$, b <u>are elements of</u> $\mathcal{D}$ <u>such that</u> $a_1b,\ldots, a_tb$ <u>form a commutative set, then the action on</u> $\mathcal{W}$ <u>of</u>

(6) $$\psi((a_1E_{j+1,j})\ldots(a_tE_{j+1,j})(bE_{j,j+1})^t),$$ <u>resp.</u>

(7) $$\psi((-a_1^*E_{j+1,j} + a_1E_{2n+1-j,2n-j})\ldots(-a_t^*E_{j+1,j} + a_tE_{2n+1-j,2n-j})(-b^*E_{j,j+1} + bE_{2n-j,2n+1-j})^t)$$

<u>agrees with</u> $f_t(a_1b,\ldots, a_tb)$, <u>where</u> f_t <u>is as before with</u> $\rho(a) = d\tau(a)$. Then the (d+1)-th symmetric identity for ρ, as displayed in §III.2 yields $f_{d+1}(ab,ab,\ldots, ab) = 0$, which is our desired result.

For $t = 1$ and $s\ell(n,\mathcal{D})$, we have

$\psi((aE_{j+1,j})(bE_{j,j+1}))=[aE_{j+1,j},bE_{j,j+1}] = abE_{j+1,j+1} - baE_{jj}$, whose effect on W is $d\tau(ab) = f_1(ab)$. Thus the underlined assertion holds for $t = 1$. Assuming $t \geq 1$, and that it holds for t, we see as before that, with commuting $a_1b,\ldots, a_{t+1}b$,

$$(8) \qquad \psi((a_1E_{j+1,j})\cdots(a_{t+1}E_{j+1,j})(bE_{j,j+1})^{t+1})$$
$$= \sum_{i=1}^{t+1} \psi((a_1E_{j+1,j})\cdots(a_ibE_{j+1,j+1}-ba_iE_{jj})\cdots(a_{t+1}E_{j+1,j})(bE_{j,j+1})^t)$$
$$= \sum_{i=1}^{t+1} \psi((a_ibE_{j+1,j+1} - ba_iE_{jj})\cdots\widehat{(a_iE_{j+1,j})}\cdots(bE_{j,j+1})^t)$$
$$- \sum_{\ell<i} \psi((a_\ell ba_i+ a_iba_\ell)E_{j+1,j})\cdots\widehat{(a_\ell E_{j+1,j})}\cdots\widehat{(a_iE_{j+1,j})}\cdots(bE_{j,j+1})^t).$$

Considering the effect of (8) on W, we see that the first sum acts as

$\sum_i d\tau(a_ib)\, \psi((a_1E_{j+1,j})\cdots\widehat{(a_iE_{j+1,j})}\cdots(bE_{j,j+1})^t)$, while the effect of the second is, by inductive hypothesis and the commutativity of $a_\ell ba_ib + a_iba_\ell b$ and all $a_\nu b$, $- \sum_{\ell<i} f_t(a_\ell ba_ib + a_iba_\ell b, a_1b,\ldots,\widehat{a_\ell b},\ldots, \widehat{a_ib},\ldots, a_{t+1}b)$.

By our commutativity once again, and the inductive hypothesis for the first sum, the result is

$$\sum_i d\tau(a_ib)f_t(a_1b,\ldots, \widehat{a_ib},\ldots, a_{t+1}b)$$
$$- 2 \sum_{\ell<i} f_t((a_\ell b)(a_ib), a_1b,\ldots, \widehat{a_\ell b},\ldots, \widehat{a_ib},\ldots, a_{t+1}b)$$

$= f_{t+1}(a_1b,\ldots, a_{t+1}b)$. This completes the inductive step for $\mathfrak{sl}(n,\mathcal{D})$. The argument in the involutory case is identical. We have therefore proved:

<u>PROPOSITION V.1. With L one of the central simple Lie algebras of this chapter, there is a unique irreducible representation of L of highest weight $d\lambda_j$ whenever i) $1 < j < n-1$ if $L = \mathfrak{sl}(n,\mathcal{D})$ or if L is associated with an involutorial algebra where $\mathcal{D}$ is quaternionic with standard involution; ii) $1 < j < n$, in the other involutorial cases. This representation is absolutely irreducible. The corresponding module is obtained from $W \otimes_{U(P)} U(L)$, where W is a one-dimensional P-module annihilated by N and with the action of L_0 as in Lemma 3 (with $k = d$), by forming the quotient of this $U(L)$-module by the submodule generated by all $w \otimes e_{-\alpha_i}$, $i \neq j$, $w \in W$, and by all $w \otimes e_{-\alpha_j}^{d+1}$, where $e_\beta \in L_\beta$.</u>

Actually the one case from group A) that has been omitted above also occurs in group F) when $d = 2$. It will be treated under that case.

3. WEIGHTS OF GROUP B.

We first assume that: $n > 2$ in the case of $s\ell(n,\mathcal{D})$, or in the involutorial case where $\mathcal{D}$ is quaternionic with standard involution; $n > 1$ in other involutorial cases. If our proposed highest weight λ is $k\lambda_1$, and if $\mathcal{W}$ is a λ-admissible irreducible L_0-module, then by Lemma 2, the action on $\mathcal{W}$ of the general element (1) or (2) of L_0 agrees with $a_1E_{11} + \tau(a_2 +\ldots+ a_n)E_{22}$ resp. $(-a_1^*E_{11} + a_1E_{2n,2n})$. In the former case, since $a_1 = -\tau(a_2 +\ldots+ a_n)1 + (a_1)_0$, $(a_1)_0 \in [\mathcal{D}\mathcal{D}]$, and since $(a_1)_0E_{22}$ annihilates $\mathcal{W}$, the action is that of $a_1E_{11} - a_1E_{22}$.

Assuming $\mathcal{W}$ is such a module, with $1 \le k \le d$, denote by $\rho(a)$, for $a \in \mathcal{D}$, the transformation of $\mathcal{W}$ representing $-aE_{11} + aE_{22}$ resp. $-a^*E_{11} + aE_{2n,2n}$. Evidently ρ is a Lie morphism of $\mathcal{D}$ into $\mathrm{End}_F(\mathcal{W})$ in the latter case. In the former, $[\rho(a), \rho(b)]$ = action of $[-aE_{11} + aE_{22} - bE_{11} + bE_{22}]$ - action of $[ab]E_{11} + [ab]E_{22}$ = (action of $[ab]E_{11}-[ab]E_{22}$) = $-\rho([ab]) = \rho([ba])$; that is, ρ is a Lie morphism of $\mathcal{D}^{opp}$ into $\mathrm{End}_F(\mathcal{W})$. In the former case, $\rho(1) = k1$; in the involutorial case, $\rho(1)$ = action of $-E_{11} + E_{2n,2n}$ = action of $-E_{11} + E_{22} - E_{2n-1,2n-1} + E_{2n,2n}$ = (action of H_1) = $k1$.

Finally, we show ρ satisfies the (k+1)-th <u>symmetric identity</u>. This follows as in §2: For $s\ell(n,\mathcal{D})$, consider the effect on $\mathcal{W}$ of

$$\psi((a_1E_{21})\ldots(a_tE_{21})(E_{12})^t) \in U(L_0),$$

an effect which must be zero if $t = k+1$. We denote this transformation of $\mathcal{W}$ by $f_t(a_1,\ldots, a_t)$, and note that $f_1(a)$ = (effect of $aE_{22} - aE_{11}$) = $\rho(a)$. In the involutorial case, we let $f_t(a_1,\ldots, a_t)$ be the effect on $\mathcal{W}$ of

$$\psi((-a_1^*E_{21} + a_1E_{2n,2n-1})\ldots(-a_t^*E_{21}+ a_tE_{2n,2n-1})(-E_{12} + E_{2n-1,2n})^t),$$

and have $f_1(a)$ = (effect of $a^*E_{22} - a^*E_{11} + aE_{2n,2n} - aE_{2n-1,2n-1}$) = (effect of $-a^*E_{11} + aE_{2n,2n}$) = $\rho(a)$. Again, $k\lambda_1$-admissibility says that f_{k+1} vanishes identically.

As in §2, it is now only a matter of establishing the basic recursion of III.1 for f_t. Thus let $a_1,\ldots, a_{t+1}$ be commuting elements of $\mathcal{D}$ (technically, of $\mathcal{D}^{opp}$, in the case of $s\ell(n,\mathcal{D})$.). Then the same computation as in §2 shows the recursion for $s\ell(n,\mathcal{D})$, and the analogous computation applies in the involutory case.

<u>Thus the highest weight-space</u> $\mathcal{W}$ <u>of an irreducible</u> L<u>-module of highest weight</u> $k\lambda_1$ (subject to our restrictions on n) <u>is an irreducible module for</u> $S^k(\mathcal{D}^{opp})$ <u>or for</u> $S^k(\mathcal{D})$, <u>according as the case is</u> $s\ell(n,\mathcal{D})$ <u>or involutorial</u>. (Actually, since $\mathcal{D} \cong \mathcal{D}^{opp}$ in the latter case, the

distinction is immaterial there.) _If_ $\rho(a)$, _for_ $a \in \mathcal{D}$, _denotes the action of_

$$\rho_k(a) = a \otimes 1 \otimes \ldots \otimes 1 + 1 \otimes a \otimes 1 \otimes \ldots \otimes 1 + \ldots + 1 \otimes \ldots \otimes 1 \otimes a$$

_on_ $\mathcal{W}$ _in the structure of_ $\mathcal{W}$ _as_ $S^k(\mathcal{D}^{opp})$ _(resp._ $S^k(\mathcal{D})$_)-module, the action of_ L_0 _on_ $\mathcal{W}$ _is given by the fact that the general element_ (1) _or_ (2) _of_ L_0 _acts as_ $-\rho(a_1)$ _resp._ $\rho(a_1)$.

For $s\ell(n,\mathcal{D})$ with $n > 2$, the highest weight $k\lambda_{n-1}$ yields analogous results, with $\rho(a)$ = action on $\mathcal{W}$ of $-aE_{n-1,n-1} + aE_{nn}$. Here it is $S^k(\mathcal{D})$ for which $\mathcal{W}$ must be an irreducible module, $\rho(a)$ is the action of $\rho_k(a)$ on this irreducible $S^k(\mathcal{D})$-module, and the general element (1) of L_0 acts on $\mathcal{W}$ by $\rho(a_n)$.

The above establish necessary conditions on $\mathcal{W}$, which we shall show to be sufficient. Those cases listed under B) and not covered by the above are $s\ell(2,\mathcal{D})$ and the involutorial case with $\mathcal{D}$ quaternionic with standard involution and $n = 2$. In the latter setting, $d = 2$, so it is really only λ_1 and $2\lambda_1$ that we are considering. The highest weight λ_1 will be treated under F), where we shall show that the Cartan product with itself of the corresponding module accounts for all irreducible modules of highest weight $2\lambda_1$. Further consideration of this case is deferred until then.

In the case $n = 2$, $1 \le k \le d$, the fundamental weights in the split case as considered in Appendix 1 are $\pi_1,\ldots,\pi_{2d-1}$, with respective restrictions to $\mathcal{T}$ of $\lambda_1, 2\lambda_1, \ldots, d\lambda_1, (d-1)\lambda_1, \ldots, \lambda_1$. Now if K is a splitting field as in §II.1 and if $\mathcal{V}$ is an irreducible L-module of highest weight $k\lambda_1$, with highest weight-space $\mathcal{W}$, the considerations of §II.1 show that $\mathcal{W}$ generates the L_K-module $\mathcal{V}_K$, and that $\mathcal{W}_K$ contains the highest weight vectors for all irreducible summands of $\mathcal{V}_K$. Thus the highest weights of all these submodules have restriction $k\lambda_1$ to $\mathcal{T}$. From our information on the π_i, each such highest weight has the form $\sum_{i=1}^{2d-1} m_i \pi_i$, where $(m_1 + m_{2d-1}) + 2(m_2 + m_{2d-2}) + \ldots + dm_d = k$. Thus if $k < d$, and if $r = m_1 + 2m_2 + \ldots + km_k$, $s = m_{2d-1} + 2m_{2d-2} + \ldots + km_{2d-k}$, we have $r + s = k$, and each irreducible L_K-module with highest weight having $k\lambda_1$ as restriction to $\mathcal{T}$ has one of the $\sum_{r+s=k} p(r)p(s)$ highest weights so obtained, _i.e._, for given $r + s = k$, there are $p(r)$ functions $\sum_{\Sigma i m_i = r} m_i \pi_i$ and $p(s)$ functions $\sum_{\Sigma i m_{2d-i} = s} m_{2d-i} \pi_{2d-i}$. Any sum of the former with one of the latter gives a (distinct) dominant integral function restricting to $k\lambda_1$, and only such sums have this restriction. If $k = d$, one has the

$\sum_{0 \le r \le d} p(r)p(d-r)$ highest weights obtained as above, the highest weight π_d being counted twice in the process.

That is, every irreducible L_K-submodule of V_K has one of the highest weights just listed. We next construct a finite family of irreducible L-modules $\{U_i\}$, of highest weight $k\lambda_1$, such that every irreducible L_K-module with highest weight among the above occurs in some $(U_i)_K$. Then it follows by iii) of Proposition II.2 that the $\{U_i\}$ are all the irreducible L-modules of highest weight $k\lambda_1$.

The U_i are constructed as follows, for $r + s = k$: Let X be an irreducible $S^r(\mathcal{D}^{opp})$-module, Y an irreducible $S^s(\mathcal{D})$-module. Let $a_1E_{11} + a_2E_{22} \in L_0$ $(\tau(a_1+a_2) = 0)$ act on X by $-\rho_r(a_1)$, and on Y by $\rho_s(a_2)$. Then X and Y are irreducible L_0-modules of weight $r\lambda_1$ resp. $s\lambda_1$. It will be shown below that they are <u>admissible</u>. (When $r = 0$ or $s = 0$, X or Y is simply F, with L_0 acting trivially.) By Chapter II, §2, it follows that every irreducible L_0-submodule of $X \otimes Y$ is $k\lambda_1$-admissible. These irreducible modules are our $\{U_i\}$. Assuming the X and Y to have been shown admissible for the appropriate weights, the completeness of the $\{U_i\}$ in the sense of the last paragraph can be shown as follows:

It will be recalled from §IV.5 that the minimal ideals in $S^r(\mathcal{D}^{opp})$ and in $S^s(\mathcal{D})$ are central simple algebras and from §III.3 that the image $\rho_k(A)$ generates $S^k(A)$ for any associative algebra A. It follows that X_K, Y_K are <u>isotypic</u> completely reducible $(L_0)_K$-modules, and that if X, X' (say) are nonisomorphic irreducible $S^r(\mathcal{D}^{opp})$-modules, then irreducible submodules of X_K are not $(L_0)_K$-isomorphic to irreducible submodules of X'_K. Thus the $p(r)$ inequivalent irreducible $r\lambda_1$-admissible modules X arising from $S^r(\mathcal{D}^{opp})$ have as irreducible constituents, over $(L_0)_K$, $p(r)$ inequivalent irreducible modules. The corresponding irreducible constituents of the L_K-modules obtained from the irreducible L-modules of highest weight $r\lambda_1$ associated with these $S^r(\mathcal{D}^{opp})$-modules are in turn $p(r)$ inequivalent irreducible L_K-modules whose highest weights have restriction $r\lambda_1$ to T.

It will be shown in Appendix 1 that the highest weights of those L_K-modules of the last paragraph <u>must be combinations of</u> $\pi_1, \ldots, \pi_d$. The previous reasoning shows that exactly $p(r)$ such dominant integral combinations have restriction $r\lambda_1$. Thus <u>every irreducible</u> L_K-<u>module whose highest weight is a combination of</u> $\pi_1, \ldots, \pi_d$ <u>and has restriction</u> $r\lambda_1$ <u>to</u> T <u>arises from</u> $S^r(\mathcal{D}^{opp})$ <u>as above</u>. Similarly, with r replaced by s, $\mathcal{D}^{opp}$ replaced by $\mathcal{D}$, $\{\pi_1, \ldots, \pi_d\}$ by $\{\pi_d, \pi_{d+1}, \ldots, \pi_{2d-1}\}$. Now if X is of the first type, Y of the second type, V_X and V_Y the corresponding irreducible L-modules, and if $(V_X)_K$, $(V_Y)_K$ are associated with dominant integral functions μ resp.

ν on H_K, $(X \otimes Y)_K = X_K \otimes_K Y_K$ contains a vector of weight $\mu + \nu$ for H_K, the highest weight that can occur in $(V_X)_K \otimes_K (V_Y)_K$.

Thus the irreducible representation of L_K with highest weight $\mu + \nu$ occurs among the irreducible L_K-modules resulting from irreducible L-modules V_W, W an irreducible L_0-submodule of $X \otimes Y$. Our earlier observations show that every dominant integral function π on H_K whose restriction to T is $k\lambda_1$ can be expressed as $\pi = \mu + \nu$ for some integers r,s with $r + s = k$, where $\mu|T = r\lambda_1$, $\nu|T = s\lambda_1$, and where μ is a combination of $\pi_1,\ldots,\pi_d$, and ν a combination of $\pi_d,\ldots,\pi_{2d-1}$. In other words, every irreducible L_K-module whose highest weight has restriction $k\lambda_1$ to T occurs as a constituent of some $(V_W)_K$, where W is an irreducible L_0-constituent of some $X \otimes Y$. This proves that such modules W are the complete set of $k\lambda_1$-admissible irreducible L_0-modules $(k \le d)$.

(In fact, the central simplicity of minimal ideals for $S^r(\mathcal{D}^{opp})$ and $S^s(\mathcal{D})$ implies that $X \otimes Y$ and $(X \otimes Y)_K$ are <u>isotypic</u> modules for $S^r(\mathcal{D}^{opp}) \otimes_F S^s(\mathcal{D}) = \mathcal{B}$ resp. $\mathcal{B}_K$. One sees easily that the elements $\rho_r(a_1) \otimes 1 + 1 \otimes \rho_s(a_2)$, where $a_1E_{11} + a_2E_{22} \in L_0$, generate $\mathcal{B}$, so that these modules are also isotypic for L_0 resp. L_{0_K}. Thus each pair X,Y gives rise to a <u>unique</u> irreducible L-module and a unique irreducible L_K-module. From examination of the weights, we see that the resulting L-modules (or L_K-modules) for X,Y and for X', Y' are isomorphic only if $X \cong X'$ and $Y \cong Y'$, except that when $r = d$, $X = F$ with $a_1E_{11} + a_2E_{22}$ acting by $-d\tau(a_1)$, we have isomorphism with $s = d$, $Y' = F$ with $a_1E_{11} + a_2E_{22}$ acting by $d\tau(a_2)$ $[= -d\tau(a_1)]$. This duplication corresponds to the repetition of π_d in our consideration of weights for the split case.)

It remains to show admissibility for our L_0-modules. Assuming $n > 2$, we consider the modules W claimed to be $k\lambda_{n-1}$-admissible for $\mathfrak{sl}(n,\mathcal{D})$, $1 \le k \le d$, namely those where the general element (1) of L_0 acts as does $\rho_k(a_n) \in S^k(\mathcal{D})$ on an irreducible right $S^k(\mathcal{D})$-module W. That the definition results in an irreducible L_0-module follows from the generation of $S^k(\mathcal{D})$ by the elements $\rho_k(a)$, $a \in \mathcal{D}$. It is immediate from $\rho_k(1) = k1$ that the module has weight $k\lambda_{n-1}$. For $j < n-1$, the typical $\psi(e_{-\alpha_j} e_{\alpha_j})$ is equal to $[aE_{j+1,j}, bE_{j,j+1}] = abE_{j+1,j+1} - baE_{jj}$ for $a,b \in \mathcal{D}$, and this annihilates W. The final test is the annihilation of W by all $\psi(e_{-\alpha_{n-1}}^{k+1} e_{\alpha_{n-1}}^{k+1})$, i.e., by all $\psi((aE_{n,n-1})^{k+1}(bE_{n-1,n})^{k+1})$.

Acting in W, $\psi((aE_{n,n-1})(bE_{n-1,n})) = abE_{nn} - baE_{n-1,n-1}$ coincides with $\rho_k(ab)$. As in §2, we show that if $a_1,\ldots,a_t \in \mathcal{D}$, $b \in \mathcal{D}$, and if $\{a_1b,\ldots,a_tb\}$ is a commutative set, then the action of

(9) $$\psi((a_1E_{n,n-1})\dots(a_tE_{n,n-1})(bE_{n-1,n})^t)$$

agrees with $f_t(a_1b,\dots, a_tb)$, f_t formed from ρ_k according to (2) of Chapter III. Then, because ρ_k satisfies the (k+1)-th symmetric identity, we have the annihilation of $\mathcal{W}$ by all elements (9) with $t = k + 1$, at least for $a_1 = \dots = a_{k+1}$, and $\mathcal{W}$ is $k\lambda_{n-1}$-admissible.

Assuming the action of (9) agrees with f_t (here $t \geq 1$, and we have shown the result for $t = 1$) for $a_1,\dots, a_t$, b as specified, let $\{a_1b,\dots, a_{t+1}b\}$ be a commutative set. Then

$$\psi((a_1E_{n,n-1})\dots(a_{t+1}E_{n,n-1})(bE_{n-1,n})^{t+1})$$

$$= \sum_{i=1}^{t+1} \psi((a_1E_{n,n-1})\dots(a_ibE_{nn} - ba_iE_{n-1,n-1})\dots(bE_{n-1,n})^t)$$

$$= \sum_{i=1}^{t+1} \psi((a_ibE_{nn} - ba_iE_{n-1,n-1})\dots\widehat{(a_iE_{n,n-1})}\dots(bE_{n-1,n})^t)$$

$$- \sum_{\ell<i} \psi((a_\ell ba_i + a_iba_\ell)E_{n,n-1})\dots\widehat{(a_\ell E_{n,n-1})}\dots\widehat{(a_iE_{n,n-1})}\dots(bE_{n-1,n})^t)$$

Now the summands in the decomposition of $\mathcal{U}(L)$ defining ψ are stable under left multiplication by $\mathcal{U}(L_0)$, so the terms in the first sum above are the same as

$$(a_ibE_{nn} - ba_iE_{n-1,n-1})\psi((a_1E_{n,n-1})\dots\widehat{(a_iE_{n,n-1})}\dots(bE_{n-1,n})^t)$$

Now by induction and our commutativity hypothesis, we see as in §2 that the action of our element is

$$\sum_i \rho(a_ib)f_t(a_1b,\dots, \widehat{a_ib},\dots, a_{t+1}b)$$

$$- 2 \sum_{\ell<i} f_t((a_\ell b)(a_ib), a_1b,\dots, \widehat{a_\ell b},\dots, \widehat{a_ib},\dots, a_{t+1}b)$$

$= f_{t+1}(a_1b,\dots, a_{t+1}b)$. This completes the induction and the proof of $k\lambda_{n-1}$-admissibility.

The argument in all other cases is essentially the same. We illustrate only for $\mathfrak{sl}(2,\mathcal{D})$, where $r \leq d$, with $\rho_r:\mathcal{D} \to S^r(\mathcal{D}^{opp})$, $\mathcal{W}$ an irreducible right $S^r(\mathcal{D}^{opp})$-module, and $a_1E_{11} + a_2E_{22} \in L_0$ being represented by the action of $-\rho_r(a_1)$ on $\mathcal{W}$: Again it is straightforward that $\mathcal{W}$ is an irreducible L_0-module of weight $r\lambda_1$. The action of $\psi((aE_{21})(bE_{12}))$ $= abE_{22} - baE_{11}$ is the same as that of $\rho_r(ba)$. Now an argument like that above shows that if $t \geq 1$, and if $a_1,\dots, a_t$, $b \in \mathcal{D}$ with $\{ba_1,\dots, ba_t\}$ a commutative set, then the action of $\psi((a_1E_{11})\dots(a_tE_{21})(bE_{12})^t)$ on $\mathcal{W}$ agrees with $f_t(ba_1,\dots, ba_t)$, f_t defined by the same recursion from the action of ρ_r. Then the $r\lambda_1$-admissibility of $\mathcal{W}$ is a consequence of the (r+1)-th symmetric identity for ρ_r.

The propositions below bring together the results of this section:

PROPOSITION V.2. The irreducible modules of highest weight $k\lambda_1$, $1 \le k \le d$, for $s\ell(n,\mathcal{D})$ with $n > 2$, are constructed as follows: Let $\mathcal{W}$ be an irreducible right module for $S^k(\mathcal{D}^{opp})$; then $\mathcal{W}$ becomes a $k\lambda_1$-admissible right L_0-module with $\sum_{i=1}^{n} a_i E_{ii} \in L_0$ acting by $-\rho_k(a_1)$. The quotient of $\mathcal{W} \otimes_{\mathcal{U}(P)} \mathcal{U}(L)$ by the submodule generated by all $w \otimes e_{-\alpha_j}$, $j > 1$, and by all $w \otimes e_{-\alpha_1}^{k+1}$, is a finite-dimensional irreducible L-module of highest weight $k\lambda_1$. There are $p(k)$ nonisomorphic such modules.

PROPOSITION V.2'. Let $n > 2$ for $s\ell(n,\mathcal{D})$ as in Proposition 2. The irreducible modules of highest weight $k\lambda_{n-1}$, $1 \le k \le d$, are constructed from irreducible right $S^k(\mathcal{D})$-modules $\mathcal{W}$ by letting $\Sigma a_i E_{ii}$ act in $\mathcal{W}$ as does $\rho_k(a_n)$ to obtain a $k\lambda_{n-1}$-admissible irreducible L_0-module $\mathcal{W}$, then factoring $\mathcal{W} \otimes_{\mathcal{U}(P)} \mathcal{U}(L)$ by the submodule generated by all $w \otimes e_{-\alpha_j}$, $j < n-1$, and by all $w \otimes e_{-\alpha_{n-1}}^{k+1}$. There are $p(k)$ nonisomorphic such modules.

PROPOSITION V.3. Let $\mathcal{D}$ be an involutorial central division algebra (of first kind), $[\mathcal{D}:F] = d^2 \ge 4$. Let L be the Lie algebra of skew $\mathcal{D}$-endomorphisms of a $\mathcal{D}$-vector space of dimension $2n$, carrying a non-degenerate antihermitian form of Witt index n. Assume $n \ge 1$, and that $n \ge 3$ if $\mathcal{D}$ is quaternionic with standard involution. Let T and its roots be labeled as in §1. Then the irreducible L-modules of highest weight $k\lambda_1$, $1 \le k \le d$, are constructed as follows: Let $\mathcal{W}$ be an irreducible right module for $S^k(\mathcal{D})$, and make $\mathcal{W}$ into an L_0-module by having $\sum_{i=1}^{n} (-a_i^* E_{ii} + a_i E_{2n+1-i,2n+1-i})$ act by $\rho_k(a_1)$. Then $\mathcal{W}$ is $k\lambda_1$-admissible and one follows the procedures of Proposition 2 to construct the $p(k)$ non-isomorphic irreducible L-modules of this highest weight.

PROPOSITION V.4. The irreducible modules of highest weight $k\lambda_1$, $1 \le k \le d$, for $s\ell(2,\mathcal{D})$ are constructed as follows: For each pair of non-negative integers r,s with $r + s = k$, let X be an irreducible $S^r(\mathcal{D}^{opp})$ module and Y an irreducible $S^s(\mathcal{D})$-module. Make X and Y into L_0-modules as in Proposition 2 resp. 2'. Let $\mathcal{W}$ be an irreducible L_0-submodule of $X \otimes Y$. Then $\mathcal{W}$ is $k\lambda_1$-admissible, and the quotient of $\mathcal{W} \otimes_{\mathcal{U}(P)} \mathcal{U}(L)$ by the submodule generated by all $w \otimes e_{-\alpha_1}^{k+1} = w \otimes (aE_{21})^{k+1}$ $(a \in \mathcal{D})$ is an irreducible L-module of highest weight $k\lambda_1$. There are $\sum_{r=0}^{k} p(r)p(k-r)$ nonisomorphic

<u>modules</u> <u>if</u> $k < d$, $\sum_{r=0}^{d} p(r)p(d-r) - 1$ <u>modules</u> <u>for</u> $k = d$.

4. <u>WEIGHTS</u> <u>OF</u> <u>GROUP</u> C.

We first assume $1 < j < n-2$ in the case of $\mathfrak{sl}(n,\mathcal{D})$, $1 < j < n-1$ in the involutorial case, with $1 < j < n-2$ if $\mathcal{D}$ is quaternionic with standard involution. Then the considerations of Lemmas 1 and 2 apply to the weight $\lambda = (d-k)\lambda_j + k\lambda_{j+1}$ to yield that if $\mathcal{W}$ is a λ-admissible irreducible L_0-module, the effect on $\mathcal{W}$ of the general element (1) or (2) of L_0 agrees with that of

$$\tau(a_1 +\ldots+ a_j)E_{jj} + a_{j+1}E_{j+1,j+1} + \tau(a_{j+2}+\ldots+ a_n)E_{j+2,j+2}$$

resp.

$$\tau(a_1 +\ldots+ a_j)(E_{2n+1-j,2n+1-j} - E_{jj}) + (-a^*_{j+1} E_{j+1,j+1} + a_{j+1}E_{2n-j,2n-j}).$$

In the former case, let $\sigma(a)$ denote the effect on $\mathcal{W}$ of $-aE_{j+1,j+1} + aE_{j+2,j+2}$, $\rho(a)$ the effect of $-aE_{jj} + aE_{j+1,j+1}$. Then $\rho(a) + \sigma(a)$ is the action of $-\tau(a)E_{jj} + \tau(a)E_{j+2,j+2} = \tau(a)(H_j + H_{j+1})$, which is scalar multiplication by $d\tau(a)$. In the involutorial case, let $\rho(a)$ be the action of

$$- a^*E_{j+1,j+1} + a^*E_{j+2,j+2} - aE_{2n-j-1,2n-j-1} + aE_{2n-j,2n-j},$$

$\sigma(a)$ that of

$$- a^*E_{jj} + a^*E_{j+1,j+1} - aE_{2n-j,2n-j} + aE_{2n+1-j,2n+1-j} ,$$

so that $\rho(a) + \sigma(a)$ = effect of $\tau(a)(E_{2n+1-j,2n+1-j} - E_{jj})$ = effect of $\tau(a)\ (H_j + H_{j+1}) = d\tau(a)$, again.

The effect of $\psi((aE_{j+2,j+1})(E_{j+1,j+2})) \in U(L_0)$ in the former case is $\sigma(a)$; in the latter, $\rho(a)$ is the effect of $\psi((-a^*E_{j+2,j+1} + aE_{2n-j,2n-j-1})(-E_{j+1,j+2} + E_{2n-j-1,2n-j}))$, as in §3, and likewise for $\rho(a)$ resp. $\sigma(a)$ with j replaced by $j-1$. Now it follows from admissibility as in §3 that, in the involutorial case, ρ <u>satisfies</u> <u>the</u> $(k+1)$-th <u>symmetric</u> <u>identity</u>, $\sigma(a) = d\tau(a) - \rho(a)$ <u>the</u> $(d-k+1)$-th <u>symmetric</u> <u>identity</u>. We have $\rho(1) = k1$, and $\rho([ab]) = [\rho(a), \rho(b)]$ as before. Now Proposition IV.2 applies to show that $\rho(a)$ is the image of $\rho_k(a) \in S^k(\mathcal{D})$ <u>in the unique irreducible representation of</u> $S^k(\mathcal{D})$ <u>of</u> §IV.2, <u>i.e.</u>, the irreducible representation whose kernel is the ideal M of that section, and in which the image of $S^k(\mathcal{D})$ is the central simple algebra $\mathcal{B}$, $[\mathcal{B}:F] = \binom{d}{k}^2$.

When the algebra L is $\mathfrak{sl}(n,\mathcal{D})$, $\rho(a)$ is the image of $\rho_{d-k}(a) \in S^{d-k}(\mathcal{D})$ in the corresponding unique representation.

Thus there is at most one irreducible λ-admissible L_0-module, and it is given by requiring that the general element (1) or (2) of L_0 act on the irreducible $S^{d-k}(\mathcal{D})$-module W, (resp. $S^k(\mathcal{D})$-module W) of the last paragraph by

$$(10) \qquad -\sigma(a_1 + \ldots + a_{j+1}) - \rho(a_1 + \ldots + a_j), \quad \text{resp.}$$
$$\rho(a_{j+1}) + d\tau(a_1 + \ldots + a_j).$$

(That these are the actions of the first paragraph of this section is seen by recalling that $\tau(\sum_{i=1}^{n} a_i) = 0$ in the case of $\mathfrak{sl}(n,\mathcal{D})$, and that

$E_{2n+1-j,2n+1-j} - E_{jj} = H_j + H_{j+1} + (-E_{j+2,j+2} + E_{2n-1-j,2n-1-j})$, which acts by multiplication by d, in the involutorial case.)

Now if we define an action of L_0 on the irreducible $S^{d-k}(\mathcal{D})$-module W as in (10), we have that the action on W of $[\Sigma a_i E_{ii}, \Sigma b_i E_{ii}] = \Sigma [a_i b_i] E_{ii}$ in the case of $\mathfrak{sl}(n,\mathcal{D})$ is $-\sigma(\sum_{i=1}^{j+1} [a_i b_i]) - \rho(\sum_{i=1}^{j} [a_i b_i])$

$= -\sigma([a_{j+1} b_{j+1}]) - d\tau(\sum_{i=1}^{j} [a_i b_i]) = -\sigma([a_{j+1}, b_{j+1}])$

$= [\sigma(a_{j+1}), \sigma(b_{j+1})]$, since $\sigma = d\tau - \rho$ is a Lie anti-homomorphism. On the other hand, $[-\sigma(\sum_{i=1}^{j+1} a_i) - \rho(\sum_{i=1}^{j} a_i), -\sigma(\sum_{i=1}^{j+1} b_i) - \rho(\sum_{i=1}^{j} b_i)]$

$= [-\sigma(a_{j+1}) - d\tau(\sum_{i=1}^{j} a_i), \ -\sigma(b_{j+1}) - d\tau(\sum_{i=1}^{j} b_i)]$

$= [\sigma(a_{j+1}), \sigma(b_{j+1})]$, so W is an L_0-module. A similar verification applies in the other case. From (10) we see that W has weight λ : H_i annihilates W if $i \neq j, j+1$; in the case of $\mathfrak{sl}(n,\mathcal{D})$,

$H_j = -E_{jj} + E_{j+1,j+1}$ acts by $\rho(1) = (d-k)1$, and

$H_{j+1} = -E_{j+1,j+1} + E_{j+2,j+2}$ by $\sigma(1) = k \cdot 1$. In the involutorial case,

$H_j = -E_{jj} + E_{j+1,j+1} - E_{2n-j,2n-j} + E_{2n+1-j,2n+1-j}$ acts by

$-\rho(1) + d\tau(1) = d-k$, and H_{j+1} by $\sigma(1) = k$. Evidently any L_0-submodule of W is an $S^{d-k}(\mathcal{D})$ resp. $S^k(\mathcal{D})$-submodule (using $\sigma = d\tau - \rho$), so that W is an irreducible L_0-module of weight λ.

Finally, to see that W is λ-admissible, it follows at once that, if $i \neq j, j+1$, W is annihilated by $\psi(e_{-\alpha_i} e_{\alpha_i}) = [e_{-\alpha_i}, e_{\alpha_i}]$ for all $e_{-\alpha_i} \in L_{-\alpha_i}$, $e_{\alpha_i} \in L_{\alpha_i}$. For $\mathfrak{sl}(n,\mathcal{D})$, we have that

$\psi((aE_{j+1,j})(bE_{j,j+1})) = abE_{j+1,j+1} - baE_{jj}$ acts by $-\sigma(ab-ba) + \rho(ba)$ $= \rho(ab-ba) - d\tau(ab-ba) + \rho(ba) = \rho(ab)$, i.e., by the action of $\rho_{d-k}(ab) \in S^{d-k}(\mathcal{D})$. Now the argument of §3 applies to show that the action of $\psi((a_1E_{j+1,j})\dots(a_tE_{j+1,j})(bE_{j,j+1})^t)$ is given by $f_t(a_1b,\dots, a_tb)$ where the arguments a_ib are assumed to commute, and where the sequence $f_1,\dots,f_t,\dots$ is generated from $f_1 = \rho$ by the fundamental recursion (1) of Chapter III. In particular, when $t = d - k + 1$ and $a_1 =\dots= a_{d-k+1}$ we have $\psi((aE_{j+1,j})^{d-k+1}(bE_{j,j+1})^{d-k+1})$ acting as does $f_{d-k+1}(ab,\dots, ab) = 0$. Thus the sufficient condition for λ-admissibility is satisfied here. The remaining cases are completely analogous; for example, that

$$\psi((-a^*E_{j+1,j} + aE_{2n+1-j,2n-j})^{d-k+1}(-b^*E_{j,j+1}+ bE_{2n-j,2n+1-j})^{d-k+1})$$

annihilates $\mathcal{W}$ in the involutorial case uses that

$$[-a^*E_{j+1,j} + aE_{2n+1-j,2n-j}, -b^*E_{j,j+1} + bE_{2n-j,2n+1-j}]$$

acts by $-\rho(ba) + d\tau(ab) = d\tau(ba) - \rho(ba) = \sigma(ba)$ and that σ satisfies the $d - k + 1$-th symmetric identity. We have thus displayed <u>the unqiue irreducible</u> λ-<u>admissible</u> L_0-<u>module</u> $\mathcal{W}$, under the restrictions at the beginning of the section.

To treat the remaining cases $(d-k)\lambda_1 + k\lambda_2$, and for $\mathfrak{sl}(n,\mathcal{D})$ the case $(d-k)\lambda_{n-2} + k\lambda_{n-1}$, we have $n \geq 3$ in all cases. First consider $\mathfrak{sl}(n,\mathcal{D})$ for $n \geq 3$. Then the verification above shows that the irreducible $S^{d-k}(\mathcal{D})$-module $\mathcal{W}$ with $j = 1$ resp. $j = n-2$ in (10) is an irreducible λ-admissible L_0-module for $\lambda = (d-k)\lambda_1 + k\lambda_2$ resp. $\lambda = (d-k)\lambda_{n-2} + k\lambda_{n-1}$. On the other hand, if V is any irreducible L-module of highest weight λ, the extension V_K has $(V_K)^+ = (V^+)_K$, the subspace annihilated by N, and this necessarily contains the highest weight-spaces of V_K. Thus if π is the highest weight of any irreducible constituent of V_K, $\pi|T = \lambda$. The determinations in Appendix 1 show that only <u>one</u> dominant integral function π has this property, <u>viz</u>. $\pi = \pi_{d+k}$ resp. $\pi_{(n-2)d+k}$. It follows that all the irreducible constituents of V_K are isomorphic, and are isomorphic to irreducible constituents of the irreducible L-module of highest weight λ constructed from $\mathcal{W}$. By Proposition 2 of Chapter II, V is isomorphic to the latter module. Thus $V^+ \cong_{L_0} \mathcal{W}$, and $\mathcal{W}$ is the <u>unique</u> λ-admissible irreducible L_0-module.

Likewise in the involutorial case, the irreducible $S^k(\mathcal{D})$-module $\mathcal{W}$ with $j = 1$ in (10) is again an irreducible λ-admissible L_0-module, and is unique as above, since once again only $\pi = \pi_{d+k}$ restricts to λ among dominant integral functions on H_K. A similar argument applies to the case $\lambda = (d-k)\lambda_{n-2} + k\lambda_{n-1} = \lambda_{n-2} + \lambda_{n-1}$ that was previously excluded when $\mathcal{D}$ is quaternionic with standard involution; here $\pi = \pi_{2(n-2)+1}$ is the unique dominant integral function relative to H_K with λ as restriction.

This completes the constructions for λ in groups A), B), C) (with the quaternionic involutorial exception in some cases), and with it the fundamental representations for $\mathfrak{sl}(n,\mathcal{D})$.

5. WEIGHTS OF GROUP D.

The algebra here is involutorial, and we assume at first that $n > 1$. Lemma 1 applies to give that the action on an irreducible $\lambda(= k\lambda_n)$-admissible L_0-module W of the general element (2) of L_0 is that of $\tau(a_1 +\ldots+ a_n)(E_{n+1,n+1} - E_{nn})$, or $k\tau(a_1 +\ldots+ a_n)$. Thus if such a module for L_0 exists it is one-dimensional and unique.

One verifies as in previous cases that such a W is annihilated by all $\psi(e_{-\alpha_i} e_{\alpha_i}) \in U(L_0)$ for $i < n$. The action of $\psi(e_{-\alpha_n} e_{\alpha_n}) = [e_{-\alpha_n} e_{\alpha_n}]$ $= [aE_{n+1,n}, bE_{n,n+1}] = abE_{n+1,n+1} - baE_{nn}$ where $a^* = a$, $b^* = b$, is $k\tau(ab)$, and one sees as before that if $a_1,\ldots, a_t$, b are $*$-fixed elements of $\mathcal{D}$ such that $a_1b,\ldots, a_tb$ commute, then the action of

$$\psi((a_1E_{n+1,n})\cdots(a_tE_{n+1,n})(bE_{n,n+1})^t) \in U(L_0)$$

is $f_t(a_1b,\ldots, a_tb)$, where f_t is generated from $\rho = f_1 = k\tau$ by the basic recursion (1) of Chapter III. We are interested in $k = d$ when the involution $*$ is orthogonal and in $k = \frac{d}{2}$ when $*$ is symplectic (cases D_1, D_2, respectively).

When $k = d$, we know that $\rho = d\tau$ satisfies the $(d+1)$-th symmetric identity on all of $\mathcal{D}$ (Chapter III, §4), hence certainly on products of two $*$-fixed elements. This disposes of case D_1, for $n > 1$: there is a unique irreducible $d\lambda_n$-admissible L_0-module W; W is one-dimensional and (2) $\in L_0$ acts by $d\tau(a_1 +\ldots+ a_n)$. Actually the same conclusions hold in the case D_2, but here we must also treat $k = \frac{d}{2}$.

In fact, what must be shown is that, on products of two $*$-fixed elements in $\mathcal{D}$ ($*$ symplectic), the function $\rho = \frac{d}{2}\tau$ satisfies the $(\frac{d}{2} + 1)$-th symmetric identity. Now if $\mathcal{D}$ is of this type, $a^* = a$, $b^* = b$, we may assume neither a nor b is zero. Define $c' = b^{-1} c^*b$, for $c \in \mathcal{D}$. Then $c \mapsto c'$ is an involution in $\mathcal{D}$, and $c' = c$ if and only if $(cb^{-1})^* = cb^{-1}$, or if and only if $c = ab$, where $a^* = a$. Thus the mapping $a \mapsto ab$ sends the $*$-fixed elements onto the $'$-fixed elements, and $'$ is also of symplectic type. Thus it suffices to prove that $\rho = \frac{d}{2}\tau$ satisfies the $(\frac{d}{2} + 1)$-th symmetric identity on *-fixed elements, whenever * is an involution in $\mathcal{D}$ of symplectic type.

To show this last, we may pass to an algebraically closed extension K to assume $\mathcal{D} = \mathrm{End}_K(V)$, V a d-dimensional vector space over K. We further have $d = 2s$, and may assume the involution in $\mathcal{D}$ is the adjoint with respect to a symplectic form (u,v) on V. Thus for $T \in \mathcal{D}$,

$$(uT^*,v) = (u,vT) \quad \text{for all } u,v \in V.$$

It is shown below that if $T^* = T$, then T <u>stabilizes a pair of dual s-dimensional totally</u> isotropic subspaces of V. If U_1 and U_2 are such subspaces, then for each $u_2 \in U_2$, u_2T is the unique element x of U_2 such that for all $u_1 \in U_1$, $(u_1T,u_2) = (u_1,x)$, and is uniquely determined by $T|U_1$. If $(w_1,\dots, w_s)$ is a basis for U_1, and $(w_{2s},\dots, w_{s+1})$ the corresponding dual basis (in order) for U_2, we have

$$d\tau(T) = \mathrm{Tr}(T) = \sum_{i=1}^{s} (w_iT, w_{2s+1-i}) + (w_i,w_{2s+1-i}T)$$

$$= \sum_{i=1}^{s} 2(w_iT, w_{2s+1-i}) = 2\mathrm{Tr}(T|_{U_1}).$$

That is, $s\tau(T) = \frac{d}{2}\tau(T) = \mathrm{Tr}(T|_{U_1})$. Since U_1 has dimension s, we know by §4 of Chapter III that $\mathrm{Tr}_{U_1}(\cdot)$ satisfies the $(s+1)$-th symmetric identity on all endomorphisms of U_1, which is the desired result.

To see the underlined assertion, note first that we may assume V is not the "orthogonal" sum of non-singular proper T-stable subspaces. For if the claim holds for such "indecomposable" V, the result for general V follows by taking U_1 and U_2 to be sums of subspaces satisfying the same conditions in a set of "orthogonally T-indecomposable" V_i whose direct sum is V. Assuming V is indecomposable in this sense let $W \neq 0$ be a T-stable subspace which is a direct summand of V and not a direct sum of proper T-stable subspaces (with no reference to the form), and of maximal dimension with this property. Then W is cyclic, generated by a $w \neq 0$ with $w(T-\lambda I)^m = 0$ for some $\lambda \in K$ with $m = \dim W$, so that $w(T - \lambda I)^{m-1} \neq 0$. Thus every element of W has the form $wp(T)$, where $p(X)$ is a polynomial in $K[X]$, so $p(T)$ is *-fixed, and

$$(wp(T), wq(T)) = (w, wq(T)p(T))$$

$$= (w, wp(T)q(T)) = (wq(T), wp(T)) = 0,$$

since the form is symplectic. Thus W <u>is totally isotropic</u>, and $W \neq V$.

Let U be a second T-indecomposable, T-stable summand of V, with $(U, w(T - \lambda I)^{m-1}) \neq 0$. (Some summand must have this property.) Then if u is a generator for U, with $u(T - \mu I)^{n-1} \neq 0$, $u(T - \mu I)^n = 0$ as for W, we must have $\mu = \lambda$; for otherwise every element of U is in the image of $T - \lambda I$, and $(x(T - \lambda I), w(T - \lambda I)^{m-1}) = (x, w(T - \lambda I)^m) = 0$. Moreover, we must have $n \leq m$ by the maximality of dimension for W. On the other hand, from the fact that $(u(T - \lambda I)^j, w(T - \lambda I)^{m-1}) = 0$ for all $j > 0$, $(u(T - \lambda I)^{m-1}, w) = (u,w(T - \lambda I)^{m-1}) \neq 0$, so $n = m$. By assumption, $U \cap W = 0$; we next note that U and W are dual with respect to (,), so that $U + W = U \oplus W$ is a nonsingular T-stable subspace of V, and must therefore be equal to V. Since W is also totally isotropic, this gives the result.

For the duality, let $y = \sum_{i=0}^{m-1} \alpha_i w(T - \lambda I)^i \neq 0$; it suffices to display $x \in U$ with $(x,y) \neq 0$. Now if j is the smallest index with $\alpha_j \neq 0$, take $x = u(T - I)^{m-1-j}$. Then $(x,y) = \alpha_j(u,w(T - I)^{m-1}) \neq 0$, and we are done. (One can in fact adjust the choice of u by combinations of the $u(T - I)^j$, $j > 0$, so that $(u(T - \lambda I)^j, w(T - \lambda I)^{m-1-k}) = \delta_{jk}$ -- cf. [10], [20].

Now let $n = 1$. When the involution is of orthogonal type, we have, from the deferral of B) in this case, all highest weights $k\lambda_1$, $1 \leq k \leq d$, to consider. As in §3 for $n > 1$, every irreducible $S^k(\mathcal{D})$-module is a $k\lambda_1$-admissible irreducible L_0-module with the effect of $-a^*E_{11} + aE_{22}$ on W defined to be that of $\rho_k(a)$. One obtains in this fashion $p(k)$ inequivalent $k\lambda_1$-admissible L_0-modules. When the ground field is (finitely) extended to a splitting field K, each irreducible $S^k(\mathcal{D})$-module becomes a sum of isomorphic irreducible $S^k(\mathcal{D}_K)$-modules, and $p(k)$ nonisomorphic $S^k(\mathcal{D}_K)$-modules are realized this way. In turn these are nonisomorphic irreducible $(L_0)_K$-modules occurring within the subspace annihilated by N of some V_K, where V is an irreducible L-module of highest weight $k\lambda_1$, with the L_0-module which is its highest weight space being one of our $p(k)$ $S^k(\mathcal{D})$-modules. Thus $p(k)$ nonisomorphic irreducible L_K-modules whose highest weight π has $\pi|T = k\lambda_1$ arise as submodules of these V_K. From Appendix 2 we see that there are only $p(k)$ dominant integral functions π on H_K with $\pi|T = k\lambda_1$. Now iii) of Proposition II.2 applies, and we conclude that <u>the irreducible $S^k(\mathcal{D})$-modules exhaust the $k\lambda_1$-admissible irreducible L_0-modules</u>.

When the involution * is of symplectic type, matters are more complicated. With $n = 1$, the restrictions to T of $\pi_1, \ldots, \pi_d$ are seen in Appendix 2 to be

$$\lambda_1,\ 2\lambda_1, \ldots,\ (d-2)\lambda_1,\ (\tfrac{d}{2} - 1)\lambda_1,\ \tfrac{d}{2}\lambda_1,$$

respectively. For $1 \leq k < \frac{d}{2} - 1$, the argument used in the orthogonal case of the last paragraph applies: There are exactly $p(k)$ inequivalent

$k\lambda_1$-admissible irreducible L_0-modules, which are the irreducible $S^k(\mathcal{D})$-modules with $-a^*E_{11} + aE_{22} \in L_0$ acting by $\rho_k(a)$.

More generally, for $1 \le k \le d-2$, one obtains in this way $p(k)$ inequivalent $k\lambda_1$-admissible irreducible L_0-modules. However, for $k \ge \frac{d}{2} - 1$, the set of dominant integral functions on H_K whose restrictions to T are equal to $k\lambda_1$ is somewhat larger. For $k = \frac{d}{2} - 1$, it consists of $p(k)$ combinations of $\pi_1, \ldots, \pi_k$, together with π_{d-1}; for $\frac{d}{2} \le k \le d-2$, of: $p(k)$ combinations of $\pi_1, \ldots, \pi_k$; sums of π_{d-1} and combinations of $\pi_1, \ldots, \pi_{k+1-d/2}$ whose sum restricts to $(k+1-\frac{d}{2})\lambda_1$, $p(k+1-\frac{d}{2})$ of them in all; sums of π_d and combinations of $\pi_1, \ldots, \pi_{k-d/2}$ whose sum restricts to $(k-\frac{d}{2})\lambda_1$, $p(k-\frac{d}{2})$ of these.

The $p(k)$ irreducible $k\lambda_1$-admissible modules arising from $S^k(\mathcal{D})$ are realized as minimal right ideals in $S^k(\mathcal{D})$. Upon extension to a splitting field K they become sums of isomorphic irreducible modules, nonisomorphic for nonisomorphic $S^k(\mathcal{D})$-modules, and all these modules for $S^k(\mathcal{D}_K)$ occur in $\otimes^k E$, where $\mathcal{D}_K = \mathrm{End}_K(E)$, $[E:K] = d$ (cf. Chapter IV). Further extending K (finitely) as necessary to split L, we realize E as a d-dimensional totally isotropic subspace in a K-vector space X of dimension 2d, carrying a non-degenerate alternate form, and L_K as the Lie algebra of skew K-endomorphisms of X with respect to this form (Appendix 2). The action of $(L_0)_K$ on X is essentially that of $(\mathcal{D}_0)_K$ on E, in that X is the direct sum of two isomorphic $(L_0)_K$-modules, each of which is a $(\mathcal{D})_K$-module isomorphic to E by the extension of the Lie morphism $a \to -a^*E_{11} + aE_{22}$ of $\mathcal{D}$ onto L_0. With our labeling of Appendix 2, this means that the irreducible L_K-module X has highest weight π_1, and the highest weights with respect to H_K of the L_K-submodules of $\otimes_K^k X$ are the only highest weights that occur in L_K-modules resulting by base field extension from L-modules of highest weight $k\lambda_1$ resulting from L_0-modules that are irreducible for $S^k(\mathcal{D})$, i.e., whose extensions occur in $\otimes_K^k E$.

All the weights, relative to H_K, of $\otimes_K^k X$ are of the form $k\pi_1 - \Sigma m_i\gamma_i$, where the m_i are nonnegative integers and the γ_i are the simple roots relative to H_K. Now $k\pi_1 = \sum_{i=1}^{d-2} k\gamma_i + \frac{k}{2}\gamma_{d-1} + \frac{k}{2}\gamma_d$, while

for $k < j \le d-2$, $\pi_j = \sum_{i=1}^{j} i\gamma_i + \sum_{i=j+1}^{d-2} j\gamma_i + \frac{j}{2}\gamma_{d-1} + \frac{j}{2}\gamma_d$, while

$$\pi_{d-1} = \frac{1}{2}\left(\sum_{i=1}^{d-2} i\gamma_i + \frac{d}{2}\gamma_{d-1} + \frac{d-2}{2}\gamma_d\right), \quad \pi_d = \frac{1}{2}\left(\sum_{i=1}^{d-2} i\gamma_i + \frac{d-2}{2}\gamma_{d-1} + \frac{d}{2}\gamma_d\right).$$

From this it is clear that no π_j, for $j > k$, can occur as a weight of $\otimes_K^k X$.

Thus the highest weights with respect to H_K that can occur in modules V_K, where V is an irreducible L-module of highest weight $k\lambda_1$, constructed from a minimal right ideal in $S^k(\mathcal{D})$ as above, are nonnegative integral combinations of $\pi_1,\ldots,\pi_k$, and one has (see Appendix 2) that exactly $p(k)$ such combinations have restriction $k\lambda_1$. Thus <u>all irreducible L_K-modules, and only these, whose highest weight is a combination of $\pi_1,\ldots,\pi_k$ with restriction $k\lambda_1$ to T, arise from the construction based on $S^k(\mathcal{D})$</u>. (Here $\frac{d}{2} - 1 \le k \le d-2$).

We also have $\pi_{d-1}|T = (\frac{d}{2} - 1)\lambda_1$, and $\pi_{d-1} + \sum_{j=1}^{k-\frac{d}{2}+1} m_j\pi_j$, $\sum_{j=1}^{k-\frac{d}{2}+1} jm_j = k - \frac{d}{2} + 1$, has restriction $k\lambda_1$ to T. Likewise, $\pi_d|T = \frac{d}{2}\lambda_1$ and each $\pi_d + \sum_{j=1}^{k-\frac{d}{2}+1} m_j\pi_j$, $\Sigma jm_j = k - \frac{d}{2}$, has restriction $k\lambda_1$ to T. These give $p(k - \frac{d}{2} + 1) + p(k - \frac{d}{2})$ additional dominant integral functions on H_K whose restrictions to T are equal to $k\lambda_1$; together with the $p(k)$ functions of the previous paragraph, they are exhaustive for $k < d-2$. For $k = d-2$, there is also $2\pi_{d-1}$ to consider.

First let $k = \frac{d}{2} - 1 = s-1$. Let $\mathcal{D}_1$ have the underlying vector space of $\mathcal{D}$, and let $-a^*E_{11} + aE_{22} \in L_0$ $(a \in \mathcal{D})$ act on $\mathcal{D}_1$ by right multiplication by $s\tau(a) - a^*$ in $\mathcal{D}$. Then $\mathcal{D}_1$ is an L_0-module, clearly irreducible, on which $H_1 = -E_{11} + E_{22}$ acts by scalar multiplication by $s-1$, thus an irreducible L_0-module of weight $(s-1)\lambda_1$. If $a^* = a$ and $b^* = b$, the action on $\mathcal{D}$ of $\psi((aE_{21})(bE_{12})) \in U(L_0)$ is that of $abE_{22} - baE_{11}$, or right multiplication by $s\tau(ab) + ab$ in $\mathcal{D}$.

If we denote by $\rho(a)$ the operation of right multiplication by $s\tau(a) - a^*$ on $\mathcal{D}$, for each $a \in \mathcal{D}$, <u>i.e.</u>, the action of $-a^*E_{11} + aE_{22}$ on $\mathcal{D}_1$, then it follows as in §3 that if $a_1,\ldots,a_t,b$ are *-fixed elements of $\mathcal{D}$ such that $a_1b,\ldots,a_tb$ commute, then the action on $\mathcal{D}_1$ of $\psi((a_1E_{21})\ldots(a_tE_{21})(bE_{12})^t)$ is $f_t(a_1b,\ldots,a_tb)$, where f_t is formed from ρ by the basic recursion (1) of Chapter III. In order to show $\mathcal{D}_1$ is $(s-1)\lambda_1$-admissible, it therefore suffices to show that ρ satisfies the s-th symmetric identity on products of pairs of *-fixed elements. By the argument when $n > 1$, this amounts to showing that ρ satisfies the s-th symmetric identity on fixed elements with respect to involutions of symplectic type. If $'$ is the involution $c \to c' = b^{-1}c^*b$, for $b^* = b$, we have for $x \in \mathcal{D}$, $a' = a$, $b^{-1}x\rho(a)b = b^{-1}(sx\tau(a) - xa^*)b$

$$= (b^{-1}xb)(s\tau(a) - a')$$
$$= (b^{-1}xb)(s\tau(a) - a).$$

Thus right multiplication $\rho'(a)$ by $s\tau(a) - a$ is $C_b^{-1}\rho(a)C_b$, where C_b is conjugation by b on $\mathcal{D}$. Thus it suffices to prove: $a \mapsto s\tau(a) - a \in \mathcal{D}$ satisfies the s-th symmetric identity on elements a of $\mathcal{D}$ fixed by any involution of symplectic type.

By the arguments of the case $n > 1$, we may pass to an extension field K, where the involution becomes truly symplectic in $M_d(K)$, and where the effect of a is determined by its effect on an s-dimensional totally isotropic subspace, stable under a. Here the problem reduces to showing: For every $a \in M_s(K)$, $\sigma(a) = \mathrm{Tr}(a) - a$ satisfies the s-th symmetric identity.

Now refer to Proposition IV.2, using s for d and $k = 1$. The algebra $\mathcal{B}$ in question is a homomorphic image of $S^1(A) = A = M_s(K)$ in our case, and is therefore isomorphic to $M_s(K)$; the mapping ρ is the identity, and σ is the mapping above, satisfying the s-th symmetric identity by the properties of $\mathcal{B}$. Thus $\mathcal{D}_1$ is an $(s-1)\lambda_1$-admissible irreducible L_0-module. If M is an irreducible $(k-s+1)\lambda_1$-admissible module for L_0 arising from $S^{k-s+1}(\mathcal{D})$, then any irreducible L_0-submodule of $M \otimes \mathcal{D}_1$ is $k\lambda_1$-admissible. In Appendix 2 we shall see that every irreducible L_K-summand of the module R_K where R is the irreducible L-module with R_1 as K-highest weight space, has highest weight π_{d-1}, in our previous sense.

From the considerations of $S^k(\mathcal{D})$, the irreducible L_K-modules arising from modules $M \otimes \mathcal{D}_1$ above have highest weights of the form $(k-s+1)\pi_1 + \pi_{d-1} - \Sigma\, m_i\gamma_i$, m_i nonnegative integers. The modules M are exactly the irreducible right $S^{k-s+1}(\mathcal{D})$-submodules of $\otimes^{k-s+1}(\mathcal{D})$, so those L_0-submodules of modules $M \otimes \mathcal{D}_1$ are the irreducible L_0-submodules of $(\otimes^{k-s+1}\mathcal{D}) \otimes \mathcal{D}_1$.

Now consider the mapping $\varphi\colon (\otimes^{k-s+1}\mathcal{D}) \otimes \mathcal{D}_1 \to \otimes^{k-s}\mathcal{D}$ sending $v_1 \otimes \cdots \otimes v_{k-s+1} \otimes u$ to $\tau(u^* v_{k-s+1})v_1 \otimes \cdots \otimes v_{k-s}$, where $v_1, \ldots, v_{k-s+1}, u \in \mathcal{D}$. The action of $a \in \mathcal{D}$, determined by that of $-a^*E_{11} + aE_{22} \in L_0$, on $(\otimes^{k-s+1}\mathcal{D}) \otimes \mathcal{D}_1$ sends $v_1 \otimes \cdots \otimes v_{k-s+1} \otimes u$ to

$$v_1 a \otimes v_2 \otimes \cdots \otimes v_{k-s+1} \otimes u + \cdots + v_1 \otimes \cdots \otimes v_{k-s+1}a \otimes u$$
$$+ s\tau(a)v_1 \otimes \cdots \otimes v_{k-s+1} \otimes u - v_1 \otimes \cdots \otimes v_{k-s+1} \otimes ua^*,$$

an element which is sent by φ to

$\tau(u^* v_{k-s+1})(v_1 a \otimes v_2 \otimes \ldots \otimes v_{k-s} + \ldots + v_1 \otimes \ldots \otimes v_{k-s} a)$

$+ s\tau(a)\tau(u^* v_{k-s+1})(v_1 \otimes \ldots \otimes v_{k-s})$

$+ (\tau(u^* v_{k-s+1} a) - \tau(a u^* v_{k-s+1}))(v_1 \otimes \ldots \otimes v_{k-s})$

The last term is zero. If we define a new structure of Lie $\mathcal{D}$-module $\mathcal{D}^{(k-s)}$ on $\otimes^{k-s}\mathcal{D}$ by letting $a \in \mathcal{D}$ send $v_1 \otimes \ldots \otimes v_{k-s}$ to $s\tau(a)v_1 \otimes \ldots \otimes v_{k-s} + v_1 a \otimes \ldots \otimes v_{k-s} + \ldots + v_1 \otimes \ldots \otimes v_{k-s} a$, the above shows that φ is a homomorphism of $\mathcal{D}$-modules from $(\otimes^{k-s+1}\mathcal{D}) \otimes \mathcal{D}_1$ onto $\mathcal{D}^{(k-s)}$, or a homomorphism of $k\lambda_1$-admissible L_0-modules. Thus $\mathcal{D}^{(k-s)}$ is a sum of irreducible $k\lambda_1$-admissible L_0-modules, all of which occur as submodules of $(\otimes^{k-s+1}\mathcal{D}) \otimes \mathcal{D}_1$.

Upon passage to the splitting field K, the module $(\mathcal{D}^{(k-s)})_K$ is a sum of copies of $\otimes^{k-s} K^d$, the action of $- a^* E_{11} + aE_{22}$ $(a \in M_d(K))$ sending $v_1 \otimes \ldots \otimes v_{k-s}$ $(v_i \in K^d)$ to

$v_1 a \otimes \ldots \otimes v_{k-s} + \ldots + v_1 \otimes \ldots \otimes v_{k-s} a + \frac{1}{2} \mathrm{Tr}(a) v_1 \otimes \ldots \otimes v_{k-s}$.

Evidently the irreducible constituents of $\otimes^{k-s}K^d$ are the same set as those for the usual action of $M_d(K)$ via $S^{k-s}(M_d(K))$. Thus there are $p(k-s)$ inequivalent ones, corresponding to symmetry operators, and corresponding to minimal right ideals in $S^{k-s}(\mathcal{D})$. The modules for $(L_0)_K$ have the action of a typical element of H_K,

$$(11) \qquad -\left(\sum_{i=1}^{d} \delta_i e_{ii}\right)E_{11} + (\Sigma \delta_i e_{ii})E_{22}, \quad \text{in the notation of}$$

Appendix 2, on an irreducible constituent of $\otimes^{k-s}K^d$ being multiplication by

$\frac{1}{2}(\delta_1 + \ldots + \delta_d) + \sum_{i=1}^{d-2} m_i \left(\sum_{j=1}^{i} \delta_j\right)$, where $m_1 + 2m_2 + \ldots + (d-2)m_{d-2} = k-s$.

(Here $(m_1, \ldots, m_{k-s})$ is the corresponding partition of $k-s$.) The L_K-module generated by this constituent has highest weight $\pi_d + \sum_{j=1}^{k-s} m_j \pi_j$. Thus the irreducible constituents of the irreducible L-modules generated from L_0-submodules of $\mathcal{D}^{(k-s)}$ are precisely those which, upon splitting, have highest weights of this form.

When the full L_0-module $(\otimes^{k-s+1}\mathcal{D}) \otimes \mathcal{D}_1$ is considered, we decompose $\otimes^{k-s+1}\mathcal{D}$ into irreducible submodules, the inequivalent ones corresponding to partitions of $k-s+1$, and likewise over K. If $\mathcal{Y}$ is an irreducible Lie $M_d(K)$-submodule of $\otimes^{k-s+1}(K^d)$ associated with a given partition $(m_1, \ldots, m_{k-s+1})$ of $k-s+1$, then since $(\mathcal{D}_1)_K$ is a sum of d copies of the Lie $M_d(K)$-module K^d with $a \in M_d(K)$ acting by

$s\tau(a) - a^* = \frac{1}{2}\,\mathrm{Tr}(a) - a^*$, the tensor product $\mathcal{V} \otimes K^d$, as $(L_0)_K$-module, has at least one irreducible summand whose highest weight π is the sum of the separate highest weights. Thus the value of π at the general element (11) of H_K is

$$\frac{1}{2}(\delta_1 + \ldots + \delta_d) - \delta_d + \sum_{i=1}^{k-s+1} m_i \left(\sum_{j=1}^{i} \delta_j \right),$$

so that $\pi = \pi_{d-1} + \sum_{i=1}^{k-s+1} m_i \pi_i$ (see Appendix 2).

For $k < d-2$, the above enables us to conclude that the irreducible L-modules of highest weight $k\lambda_1$ resulting as submodules of $\otimes^k \mathcal{D}$ or of $(\otimes^{k-s+1}\mathcal{D}) \otimes \mathcal{D}_1$ have as irreducible summands over K all irreducible L_K-modules whose highest weight π restricts to $k\lambda_1$ on T. By Proposition II.2, these irreducible L-modules exhaust all those of highest weight $k\lambda_1$.

When $k = d-2$, $\mathcal{D}_1 \otimes \mathcal{D}_1$ contains an irreducible $k\lambda_1$-admissible L_0-module which in turn contains, upon field extension to K, an irreducible $(L_0)_K$-module of weight $2\pi_{d-1}$ (relative to H_K). Thus the irreducible L-modules induced from irreducible L_0-submodules of

$$\otimes^k \mathcal{D},\ (\otimes^{k-s+1}\mathcal{D}) \otimes \mathcal{D}_1,\ \mathcal{D}_1 \otimes \mathcal{D}_1,$$

exhaust, by Proposition II.2, all irreducible L-modules of highest weight $k\lambda_1 = (d-2)\lambda_1$.

6. WEIGHTS OF GROUPS E AND F.

In case E_1), with $n \geq 2$, $1 \leq k \leq d$, Appendix 2 shows that there is only one dominant integral function on H_K (namely $\pi_{(n-1)d+k}$) whose restriction to T is $\lambda = (d-k)\lambda_{n-1} + k\lambda_n$. As in earlier cases, it follows from Proposition II.2 that there is at most one irreducible L-admissible L_0-module. On the other hand, nothing beyond what is in §4 is needed to show that such a module W is realized as follows:

Let B be the central simple homomorphic image of $S^k(\mathcal{D})$ of §IV.2, $[B:F] = \binom{d}{k}^2$; for $a \in \mathcal{D}$, let $\rho(a)$ be the transformation of a minimal right ideal W in B consisting of right multiplication by the canonical image in B of $\rho_k(a) \in S^k(\mathcal{D})$. Let $\sum_{j=1}^{n} (-a_j^* E_{jj} + a_j E_{2n+1-j,2n+1-j}) \in L_0$ act in W by $\rho(a_n) + d\tau(a_1 + \ldots + a_{n-1})$. Accordingly, W with this action of L_0 is the unique irreducible λ-admissible L_0-module.

For weights λ of group E_2), with $d > 2$ and $k < d-2$, the same argument as to uniqueness and the same construction of the unique λ-admissible irreducible L_0-module apply. When $d = 2$, we still must consider here the case $2\lambda_{n-1}$ (for $n \geq 2$). The weight $2\lambda_{n-1}$ is twice the weight $\lambda_{n-1} + (\frac{d}{2} - 1)\lambda_n$ of F). From Appendix 2 we see that both π_{nd-2} and $2\pi_{nd-1}$ have restriction $2\lambda_{n-1}$ to T, and that these are the only dominant integral functions on H_K with this restriction. Likewise for $d > 2$, both π_{nd-2} and $2\pi_{nd-1}$ have restriction $2\lambda_{n-1} + (d-2)\lambda_n$, and are the only such dominant integral functions on L_K. The strategy for dealing with these cases is as follows: We proceed to F), and determine the unique λ-admissible irreducible L_0-module for $\lambda = \lambda_{n-1} + (\frac{d}{2} - 1)\lambda_n$, say W. Then we show that if V is the irreducible L-module with W as highest weight space, $(V \otimes V)_K$ has irreducible L_K submodules of both highest weights π_{nd-2} and $2\pi_{nd-1}$. Then Proposition II.2 is invoked to show that $V \otimes V$ contains all irreducible L-modules of highest weight 2λ, and it follows that all irreducible 2λ-admissible L_0-modules occur as submodules of $W \otimes W$.

Thus it suffices to construct W and to show the asserted property of $(V \otimes V)_K$. The uniqueness of W follows from the fact that only π_{nd-1}, among dominant integral functions on H_K, has restriction λ to T. To demonstrate one such W, we again take the underlying space of W to be $\mathcal{D}$, the action of

$$\sum_{j=1}^{n} (-a_j^* E_{jj} + a_j E_{2n+1-j,2n+1-j}) \quad \text{on}$$

$b \in \mathcal{D}$ sending b to $-ba_n^* + \frac{d}{2}\tau(a_1 + \ldots + a_n)b$. It is readily verified that W is an irreducible L_0-module of weight λ, annihilated by $[L_{-\alpha_i} L_{\alpha_i}]$ for all $i < n-1$ and by $\psi(e^2_{-\alpha_{n-1}} e^2_{\alpha_{n-1}})$ for all $e_{\pm\alpha_{n-1}} \in L_{\pm\alpha_{n-1}}$. The final condition for λ-admissibility is the annihilation of W by all $\psi(e^s_{-\alpha_n} e^s_{\alpha_n})$, $s = \frac{d}{2}$, a condition that is clearly satisfied if $d = 2$.

As before, we consider the action on W of

$$\psi((a_1E_{n+1,n})\cdots(a_tE_{n+1,n})(bE_{n,n+1})^t), \tag{12}$$

where $a_1, \ldots, a_t, b$ are $*$-fixed in $\mathcal{D}$, and where $a_1b, \ldots, a_tb$ commute. It will suffice to show for $t = s$ that each such element annihilates W. For $t = 1$, we have $\psi((aE_{n+1,n})(bE_{n,n+1})) = abE_{n+1,n+1} - baE_{n,n}$, whose action sends $c \in \mathcal{D}$ to $-cba = -c(ab)^*$. If we let $\rho(a)$, for $a \in \mathcal{D}$, be the map of $\mathcal{D}$ into $\mathcal{D}$ sending $c \in \mathcal{D}$ to $-ca^*$, and if we let $f_t(a_1, \ldots, a_t)$ be defined for commuting elements $a_1, \ldots, a_t$ of $\mathcal{D}$ by the fundamental recursion (1) of Chapter III, with $f_1(a) = \rho(a)$, then we find as before that for $*$-fixed elements $a_1, \ldots, a_t$, b with $a_1b, \ldots, a_tb$ commuting, the action on $\mathcal{D}$ of (12) is equal to $f_t(a_1b, \ldots, a_tb)$. In particular, the action of

$\psi((aE_{n+1,n})^s(bE_{n,n+1})^s)$ is equal to $f_s(ab,\ldots, ab)$, which by (6) of Chapter III is equal to

$$\sum_{P \leftrightarrow (r_1,\ldots, r_s)} \operatorname{sgn} C(P)\ |C(P)|\ \rho(ab)^{r_1}\ \rho((ab)^2)^{r_2} \ldots \rho((ab)^s)^{r_s},$$

which is right multiplication in $\mathcal{D}$ by

$$\sum \operatorname{sgn} C(P)\ |C(P)|\ (-(ab)^*)^{r_1+2r_2+\ldots+sr_s}$$

$$= (\sum_P \operatorname{sgn} C(P)\ |C(P)|)\ (-(ab)^*)^s$$

$$= 0, \text{ since } \sum_P \operatorname{sgn} C(P)\ |C(P)| = \sum_{\sigma\in S_n} \operatorname{sgn}\sigma = 0.$$

Thus the s-th symmetric identity is satisfied by ρ on products of pairs of *-fixed elements, and $\mathcal{W}$ <u>is λ-admissible</u>.

All the irreducible L_K-modules into which the L_K-module $\dot{V}_K$ breaks up, where V is the irreducible L-module with $\mathcal{W}$ as highest weight space, must have highest weight π_{nd-1}. The subspace of $V_K \otimes V_K$ on which T acts with weight 2λ is $\mathcal{W}_K \otimes \mathcal{W}_K = (\mathcal{W} \otimes_F \mathcal{W})_K$. Now both $2\pi_{nd-1}$ and π_{nd-2} restrict to 2λ on T. If R is the irreducible L_K-module of highest weight π_{nd-1}, then $R \otimes_K R$ evidently has the irreducible module M of highest weight $2\pi_{nd-1}$ as its submodule generated by $v^+ \otimes v^+$, where v^+ is the highest weight-vector of R. In M, $v^+e_{-\gamma_{nd-1}} \otimes v^+ + v^+ \otimes v^+ e_{-\gamma_{nd-1}}$ spans the space of weight $2\pi_{nd-1} - \gamma_{nd-1} = \pi_{nd-2}$, so this weight space has dimension <u>one</u>. On the other hand, it is easy to see that π_{nd-2} has multiplicity <u>two</u> in $R \otimes R$, and therefore (as in [26], Appendix A), that the irreducible L_K-module of highest weight π_{nd-2} is present in $R \otimes_K R$, so that the highest weight-space for this module is present in $(\mathcal{W}\otimes\mathcal{W})_K$. This is enough for us to conclude that <u>all the irreducible $2\lambda_{n-1} + (d-2)\lambda_n$-admissible L_0-modules occur in $\mathcal{W} \otimes \mathcal{W}$</u>.

7. <u>SUMMARY</u>.

In the preceding four sections, all irreducible λ-admissible L_0-modules, for λ among A) - F), have been constructed. For arbitrary $\lambda = \sum_{i=1}^{n} m_i\lambda_i$, m_i nonnegative integers, the general theory of Cartan multiplication of Chapter II and the specific determination of fundamental weights for the algebras of this chapter show that the λ-admissible irreducible L_0-modules are constructed by:

1) partitioning λ into a sum of the appropriate set of fundamental weights;

2) for each such partition and each time a fundamental weight μ occurs in the partition, selecting one of the μ-admissible irreducible L_0-modules W constructed in this chapter (possibly different W's, in case the same μ is repeated and admits more than one module W);

3) decomposing into their irreducible constituents the tensor products of the sets of modules of 2).

In general, we shall say that a set of dominant integral functions μ and μ-admissible irreducible L_0-modules W is <u>fundamental</u> if the processes 1) - 3) above yield, for every $\lambda \in T^*$, all the (finite-dimensional) λ-admissible irreducible L_0-modules. We have displayed fundamental sets consisting, for $s\ell(n,\mathcal{D})$, of all weights of groups A) - C) and of all their admissible irreducible modules as constructed in §§2-4; for the other L, of skew matrices (with the involution in $\mathcal{D}$ of orthogonal type), of those weights in groups A) - C), D_1), E_1) and their admissible modules from §§2-6; for the latter when the involution in $\mathcal{D}$ is symplectic, of weights A) - C), D_2), E_2), F) and admissible modules from §§2-6. It will be clear, upon reflection, that these sets are somewhat redundant. Some of the redundancy is eliminated in the following theorems, which sum up the results of this chapter. Subsequent chapters, dealing with other simple Lie algebras, will close with corresponding theorems using the same language.

<u>THEOREM V.1.</u> <u>For</u> $n > 2$, <u>a fundamental set for</u> $s\ell(n,\mathcal{D})$ <u>consists of the sets of weights</u> $\lambda_1, \lambda_{n-1}, d\lambda_j\ (1 < j < n-1),\ (d-k)\lambda_i + k\lambda_{i+1}\ (1 \le i < n-1,\ 1 \le k < d)$ (here $[\mathcal{D}:F] = d^2$) <u>and, for each weight</u> μ <u>in the set, of a single</u> μ-<u>admissible irreducible</u> L_0-<u>module</u> $W^{(\mu)}$. <u>The underlying space of</u> $W^{(\mu)}$ <u>is as follows:</u>

$\mu = d\lambda_j : W^{(\mu)} = F.$

$\mu = (d-k)\lambda_j + k\lambda_{j+1}: W^{(\mu)}$ <u>is a certain minimal right ideal in</u> $S^k(\mathcal{D})$, <u>lying in a unique minimal 2-sided</u> $\mathcal{B}$ <u>as determined in Chapter IV</u>, §§2,3; $\mathcal{B}$ <u>is central simple over</u> F, $[\mathcal{B}:F] = \binom{d}{k}^2$.

$\mu = \lambda_1, \lambda_{n-1} : W^{(\mu)} = \mathcal{D}.$

<u>The action of</u> L_0 <u>on</u> $W^{(\mu)}$ <u>is given in</u> §§2,3,4 <u>respectively</u>.

<u>THEOREM V.2.</u> <u>A fundamental set for</u> $s\ell(2,\mathcal{D})$ <u>consists of the single weight</u> λ_1 <u>and two inequivalent</u> λ_1-<u>admissible irreducible</u> L_0-<u>modules</u> W_1, W_2. <u>The underlying space of each of these modules is</u> $\mathcal{D}$. <u>The action of</u> L_0 <u>on</u> W_1, W_2 <u>is given in</u> §3 <u>under the labels</u> X (with $r = 1$), Y (with $s = 1$).

<u>THEOREM V.3.</u> <u>Let</u> L <u>be the Lie algebra of skew</u> $\mathcal{D}$-<u>endomorphisms of a 2n-dimensional left</u> $\mathcal{D}$-<u>vector space with respect to a non-degenerate</u>

antihermitian form of Witt index n. We canonically identify L with a Lie algebra of $2n$ by $2n$ $\mathcal{D}$-matrices as in §1, and T, L_0, $\{\alpha_i\}$, $\{\lambda_i\}$ as there. When the involution in $\mathcal{D}$ is of orthogonal type, a fundamental set for L consists of the set of weights

$$\{\lambda_1,\ d\lambda_j (1 < j \le n),\ (d-k)\lambda_i + k\lambda_{i+1} (1 \le i < n,\ 1 \le k < d)\}$$

and for each weight μ in the set, of a single μ-admissible irreducible L_0-module $\mathcal{W}^{(\mu)}$. The underlying space of $\mathcal{W}^{(\mu)}$ is as in Theorem 1, that for $d\lambda_n$ being again F. The action of L_0 on $\mathcal{W}^{(\mu)}$ is given in §§2,3,4,5 (under D_1), 6 (under E_1).

For $n = 1$, only λ_1 and the corresponding module, identified with $\mathcal{D}$ as in §4, is required for our fundamental set.

THEOREM V.4. Let L and other notations be as in Theorem 3, but assume now that the involution in $\mathcal{D}$ is of symplectic type. If $d = 2$, assume further that $n > 2$. Then a fundamental set for L consists of the set of weights

$$\{\lambda_1,\ d\lambda_j\ (1 < j < n),\ \tfrac{d}{2}\lambda_n,\ (d-k)\lambda_i + k\lambda_{i+1}\ (1 \le i < n-1,$$
$$1 \le k < d),\ (d-k)\lambda_{n-1} + k\lambda_n\ (1 \le k < d-2),\ \lambda_{n-1} + (\tfrac{d}{2} - 1)\lambda_n\}\ ,$$

and, for each μ in this set, of a single μ-admissible irreducible L_0-module $\mathcal{W}^{(\mu)}$. The underlying space of $\mathcal{W}^{(\mu)}$ is as in Theorem 3, for those μ labeled as in that fundamental set. For $\mu = \frac{d}{2}\lambda_n$, the space is F, and for $\mu = \lambda_{n-1} + (\frac{d}{2} - 1)\lambda_n$, it is $\mathcal{D}$. The action of L_0 on $\mathcal{W}^{(\mu)}$ is given in §§2-6. When $d = 2 = n$, there are two inequivalent λ_1-admissible modules.

For $n = 1$ (and $d > 2$), one must take $\lambda_{n-1} + (\frac{d}{2} - 1)\lambda_n$ to mean $(\frac{d}{2} - 1)\lambda_1$. Here λ_1 and $(\frac{d}{2} - 1)\lambda_1$ are the weights for a fundamental set, with one admissible irreducible L_0-module for each. In each case, the underlying space of the module is $\mathcal{D}$, the action of L_0 being given in §3 resp. §5 ($\mathcal{D}$ denoted as $\mathcal{D}_1$).

To prove Theorems 1-4, all that remains is to show that all irreducible λ-admissible L_0-modules for the weights λ considered in §§3-6 but not listed in the theorems occur as constituents of the appropriate tensor products of those listed in the theorems.

In Theorem 1, the action of $g = \sum_{i=1}^{n} a_i E_{ii}$, $a_i \in \mathcal{D}$, $\Sigma\, \tau(a_i) = 0$, on an irreducible $k\lambda_1$-admissible module agrees with that of $a_1E_{11} - \tau(a_1)E_{22}$, and with that of $a_1E_{11} - a_1E_{22}$. Furthermore, we have seen in §3 that the action of this last element is that of $-\rho_k(a_1) \in S^k(\mathcal{D}^{opp})$ on an irreducible right module for $S^k(\mathcal{D}^{opp})$. For $k = 1$, this module is $\mathcal{D}^{opp}$ itself and the action sends b to $-b_{(opp)}a_1$, which is $-a_1b$, in the multiplication of $\mathcal{D}$. In the k-fold tensor power $\otimes^k\mathcal{D}^{opp}$, the k-fold tensor power of the representation of L_0 represents $g \in L_0$ by right multiplication by $-\rho_k(a_1)$. From the evident fact that $\otimes^k\mathcal{D}^{opp}$ contains all irreducible $S^k(\mathcal{D}^{opp})$-modules, it is now clear that all irreducible $k\lambda_1$-admissible L_0-modules occur in the k-th tensor power of the unique λ_1-admissible module $\mathcal{D}^{opp}$. The argument for $k\lambda_{n-1}$ is completely analogous, with $\mathcal{D}$ replacing $\mathcal{D}^{opp}$. The case of $\mathfrak{sl}(2,\mathcal{D})$, set forth in Theorem 2, is handled by noting that for $1 \le k \le d$, the module $\sum_{r+s=k} (\otimes^r \mathcal{W}_1) \otimes (\otimes^s \mathcal{W}_2)$ contains all irreducible $k\lambda_1$-admissible L_0-modules. This was shown in §3 by appealing to splitting information and Proposition II.2.

Theorem 3 follows exactly as did Theorem 1; here it is the tensor powers $\otimes^k\mathcal{D}$ of the unique λ_1-admissible irreducible L_0-module $\mathcal{D}$ that contain all $k\lambda_1$-admissible modules. The same applies to $k\lambda_1$-admissible modules in Theorem 4, when $d > 2$ and $n > 1$ or when $d = 2$ and $n > 2$. For $n > 2$ (for $n > 1$ if $d > 2$), the other weight that has been eliminated from our lists of §1 is $2\lambda_{n-1} + (d-2)\lambda_n$, from group E_2). In §6 we have seen that all admissible irreducible L_0-modules for this weight occur in the tensor square of the unique admissible module for $\lambda_{n-1} + (\frac{d}{2} - 1)\lambda_n$. For $d = 2$ and $n = 2$, the list of weights of §1 consists of $\lambda_1, 2\lambda_1, \lambda_2$. The module $\mathcal{W}$ of the last sentence is λ_1-admissible $(\lambda_{n-1} = \lambda_1)$ as is the module $\mathcal{D}$, in the pattern previously established.

In the labeling of our splitting, all irreducible constituents of the L_K-module resulting from $\mathcal{D}$ have highest weight π_1, while that from $\mathcal{W}$ has constituents of highest weight π_3. Now tensor products of these give irreducible L-constituents among whose highest weights, upon splitting, are included $2\pi_1, \pi_2, 2\pi_3, \pi_1 + \pi_3$ - the only ones whose restriction to T is $2\lambda_1$. It now follows, using Proposition II.2, that our assertion holds for $n > 1$. The final case of Theorem 4, that for $n = 1$ (and $d > 2$) has been established in §6.

A rather curious consequence of our results (for the case $n = 1$, say), is the following: Let ρ be a linear mapping of $\mathcal{D}$ into a ring of endomorphisms of an F-vector space V of finite dimension. Suppose that $1 \le k \le d$ and that $\rho(1) = kI$, $\rho([ab]) = [\rho(a), \rho(b)]$ for all $a,b \in \mathcal{D}$ -- then $\rho(\mathcal{D})$ acts completely reducibly. Now suppose further that ρ satisfies the (k+1)-symmetric identity <u>on products of pairs of *-fixed elements of</u> $\mathcal{D}$. Then, if the involution in $\mathcal{D}$ is of orthogonal type, ρ satisfies the (k+1)-th symmetric identity on <u>all</u> elements of $\mathcal{D}$. If the involution is of symplectic type, <u>and if</u> $k < \frac{d}{2} - 1$, the same implication holds. On the other hand, it fails for $k = \frac{d}{2} - 1$ (and for larger k); for a counter-example for $k = \frac{d}{2} - 1$ consider the module $\mathcal{D}_1$ of §6, and for $k = \frac{d}{2}$ the map $\rho(a) = \frac{d}{2}\tau(a)$.

The idea behind the assertions of the last paragraph is that each irreducible subspace of $\mathcal{V}$ must be a $k\lambda_1$-admissible $\mathcal{L}_0$-module. With the restrictions on k, each such module has been seen to be an $S^k(\mathcal{D})$-module, by comparisons in the split case, and in these the $(k+1)$-th symmetric identity holds universally. No general "rational" explanation of this fact is known to the author.

CHAPTER VI

CONSTRUCTION OF REPRESENTATIONS: TYPE C (SECOND KIND)

In this chapter we take up the Lie algebras associated with involutions of second kind, postponed from Chapter V. That is, we have a central division algebra $\mathcal{D}$ over the field Z, Z being a quadratic extension of F, with $[\mathcal{D}:Z] = d^2$. $\mathcal{D}$ is assumed to have an involution $*$ which is F-linear but not Z-linear. Since $Z^* = Z$, we have $Z = F1 + F\zeta$ where $\zeta \in Z$, $\zeta^* = -\zeta$. Then $\zeta^2 \in Z$ is fixed by $*$, so $\zeta^2 = \gamma \in F$, $\gamma \notin F^2$. This choice will be fixed throughout.

1. THE LIE ALGEBRAS AND THEIR FUNDAMENTAL WEIGHTS.

As in §V.1, let V be a left $\mathcal{D}$-vector space of dimension $2n$, carrying a nondegenerate antihermitian form of Witt index n, with $\mathcal{D}$-basis $e_1, \ldots, e_{2n}$. Let M be the set of all $\mathcal{D}$-endomorphisms T of V such that $(uT,v) + (u,vT) = 0$ for all $u, v \in V$. Then M is a Lie algebra over F; the mapping $u \to \zeta u$ is in M and evidently centralizes M. Thus M is not simple; however $L = [MM]$ is central simple.

Relative to the basis $e_1, \ldots, e_{2n}$, M is given as the set of $2n$ by $2n$ $\mathcal{D}$-matrices (a_{ij}) satisfying the conditions of §V.1:

$$a_{ij}^* = -a_{2n+1-j,\,2n+1-i}\,;$$
$$a_{i,2n+1-j}^* = a_{j,2n+1-i}\,;$$
$$a_{2n+1-i,j}^* = a_{2n+1-j,i}\,,$$

$1 \le i,\ j \le n$. If $\tau\colon \mathcal{D} \to Z$ is defined as before, and if $\tau^+, \tau^- : \mathcal{D} \to F$ are defined by $\tau(a) = \tau^+(a) + \tau^-(a)\zeta$, then L is the subalgebra of M consisting of those matrices with $\Sigma_{i=1}^n\ \tau^-(a_{ii}) = 0$.

A maximal split torus T in L consists of all diagonal F-matrices in L, so is spanned as before by all $E_{ii} - E_{2n+1-i,2n+1-i}$, $1 \le i \le n$, and has as centralizer L_0 in L the set of all diagonal $\mathcal{D}$-matrices in L. Thus the general element g of L_0 has the form

$$g = \Sigma_{i=1}^n\ (-a_i^* E_{ii} + a_i E_{2n+1-i,2n+1-i}),$$
$$\Sigma_{i=1}^n\ \tau^-(a_i) = 0,\quad a_i \in \mathcal{D}.$$

The system of roots relative to T in L is given as in §V.1. When L is split by an extension field K, the algebra L_K is simple of type A_{2nd-1}. A fundamental system of roots $\gamma_1, \ldots, \gamma_{2nd-1}$ with respect to a Cartan subalgebra containing T will be displayed in Appendix 3. These have $\gamma_{j|T} = 0$ if $d \nmid j$, while $\gamma_{di|T} = \alpha_i = \gamma_{(2n-i)d|T}$, $1 \le i \le n$, the α_i being the fundamental roots relative to T with root-spaces as in §V.1. The corresponding fundamental dominant weights $\pi_1, \ldots, \pi_{2nd-1}$ for L_K are seen from [3], Planche 1, p. 250, to be given by

$$\pi_j = \frac{1}{2nd} \left(\Sigma_{i=1}^{j} i(2nd-j)\ \gamma_i + \sum_{i=1}^{2nd-j-1} ij\ \gamma_{2nd-i} \right).$$

It follows as before that $\pi_{jd|T} = d\lambda_j = \pi_{(2nj)d|T}$, $1 \le j \le n$; $\pi_{i|T} = i\lambda_1 = \pi_{2nd-i|T}$, $1 \le i \le d$; $\pi_{i|T} = ((j+1)d-i)\lambda_j + (i-jd)\lambda_{j+1} = \pi_{2nd-i|T}$, for $1 \le j \le n$, $jd < i < (j+1)d$. By Proposition II.3, the irreducible L-modules with these restrictions as highest weights suffice to generate all irreducible L-modules by Cartan multiplication, in the more precise sense that the mode of generating exactly those with a given highest weight is known. We designate $h_1, \ldots, h_n$ in T as for relative type C_n in §V.1.

2. <u>CONSTRUCTION OF REPRESENTATIONS: WEIGHTS</u> $d\lambda_j (j > 1)$.

It will be convenient to set $A = \{a \in \mathcal{D} \mid a^* = a\}$, $B = \{b \in \mathcal{D} \mid b^* = -b\}$. Then $A = \zeta B = B\zeta$, and $B = \zeta A$. As in §V.2, any irreducible λ-admissible L_0-module W with λ as in this section-heading is annihilated by all $[e_{-\alpha_1}, e_{\alpha_1}]$, thus by all

$$abE_{22} - baE_{11} + (ba)^* E_{2n,2n} - (ab)^* E_{2n-1,2n-1},$$

for $a, b \in \mathcal{D}$. From this, as in §V.2, it follows that the action on W of g agrees with that of

$$- \tau(a_1 + a_2)^* E_{22} + \tau(a_1 + a_2) E_{2n-1,2n-1}$$
$$+ \Sigma_{i=3}^{n} \ (-a_i^* E_{ii} + a_i E_{2n+1-i,2n+1-i}) \ .$$

(Always here with $\Sigma_{i=1}^{n} \tau^-(a_i) = 0$). One deduces the counterpart of Lemma 1 of §V.2:

<u>The action on an irreducible</u> $d\lambda_j$<u>-admissible</u> L_0<u>-module</u> $(j > 1)$ <u>of the general element</u> $g \in L_0$ <u>agrees with that of</u>

$$- \tau(a_1 +\ldots+ a_j)^* E_{jj} + \tau(a_1 +\ldots+ a_j)E_{2n+1-j,2n+1-j}$$

$$+ \Sigma_{i=j+1}^{n} (-a_i^* E_{ii} + a_i E_{2n+1-i,2n+1-i}).$$

In particular, for $j = n$, the above element is

$$- \tau(a_1 +\ldots+ a_n)^* E_{nn} + \tau(a_1 +\ldots+ a_n)E_{n+1,n+1}$$
$$= \tau^+(a_1 +\ldots+ a_n)h_n,$$

since $\tau^-(a_1 +\ldots+ a_n) = 0$. Thus whenever $\lambda = k\lambda_n$ (and $n > 1$), the action of g on $\mathcal{W}$ is multiplication by the scalar $k\tau^+(a_1 +\ldots+ a_n)$. If an irreducible $k\lambda_n$-admissible module $\mathcal{W}$ exists, we therefore have $[\mathcal{W}:F] = 1$, and $\mathcal{W}$ is unique.

The existence question is settled by showing that a one-dimensional F-space $\mathcal{W}$, with the action of $g \in L_0$ defined as above, is a $k\lambda_n$-admissible module for $k = d$; we shall also show why $\mathcal{W}$ is not $k\lambda_n$-admissible for $k < d$. Since τ vanishes on $[\mathcal{D}\mathcal{D}]$, it is clear that such a $\mathcal{W}$ is an L_0-module, and h_n acts as kI, the other h_i as 0. Thus $\mathcal{W}$ is irreducible of weight $k\lambda_n$. Moreover, the calculation previously performed for $[e_{-\alpha_1}, e_{\alpha_1}]$ shows that $\mathcal{W}$ is annihilated by all $\psi(e_{-\alpha_i} e_{\alpha_i}) = [e_{-\alpha_i}, e_{\alpha_i}] \in U(L_0)$ for $i < n$. Finally, for $a^* = a$, $b^* = b$, $[aE_{n+1,n}, bE_{n,n+1}]$ acts on $\mathcal{W}$ by $k\tau^+(ab)$, and it follows by induction as before that for $a_1, \ldots, a_t, b$ all *-fixed, with $a_1 b, \ldots, a_t b$ a commutative set, the action on $\mathcal{W}$ of

$$\psi((a_1 E_{n+1,n}) \cdots (a_t E_{n+1,n})(bE_{n,n+1})^t) \in U(L_0)$$

satisfies our basic recursion (1) of Chapter III, so is equal to, for $a_1 =\ldots= a_t = a$,

$$\sum_{P \leftrightarrow (s_1,\ldots,s_t)} \operatorname{sgn}C(P)\ |C(P)|\,(k\tau^+(ab))^{s_1}(k\tau^+((ab)^2))^{s_2} \cdots ,$$

a quantity that must be identically zero for $t = k + 1$.

We may pass to an extension field K of F containing a square root $\gamma^{\frac{1}{2}}$ of $\gamma = \zeta^2$ in F, first assuming $b = 1$ in the identity to be checked. Then $e_1 = \frac{1}{2}(1 + \gamma^{-\frac{1}{2}} \otimes \zeta)$ and $e_2 = \frac{1}{2}(1 - \gamma^{-\frac{1}{2}} \otimes \zeta)$ are central orthogonal idempotents in $K \otimes_F Z \subseteq K \otimes_F \mathcal{D}$, interchanged by the extension of the involution in $\mathcal{D}$. Thus elements $b\, e_1 + c\, e_2$ of $K \otimes \mathcal{D}$, where $b, c \in \mathcal{D}$, exhaust $K \otimes \mathcal{D}$, and such an element is *-fixed if and only if $c = b^*$. For an element $b\, e_1 + b^* e_2$, let $b = \tau^+(b)1 + \tau^-(b)\zeta + b_0$; then $b^* = \tau^+(b)1 - \tau^-(b)\zeta + b_0$ and

$$b\,e_1 + b^*e_2 = \tau^+(b)1 + \tau^-(b)\ \gamma^{-\frac{1}{2}} \otimes \zeta^2 + b_0e_1 + b_0{}^*e_2$$
$$= (\ \tau^+(b) + \tau^-(b)\ \gamma^{\frac{1}{2}}) \otimes 1 + b_0e_1 + b_0{}^*e_2 \quad .$$

The first term, in $K \otimes 1$, identifies with the element $\tau(b)$ by the F-isomorphism of Z onto the subfield $F(\gamma^{\frac{1}{2}})$ of K sending ζ to $\gamma^{\frac{1}{2}}$; the last two terms are in $K \otimes [\mathcal{D}\mathcal{D}]$; there is no term in $K \otimes \zeta$.

That is, extension of τ^+ to $K \otimes \mathcal{D}$ maps our symmetric element $b\,e_1 + b^*e_2$ to $\tau^+(b) + \tau^-(b)\gamma^{\frac{1}{2}} \in K$, an element that identifies with $\tau(b)$ in Z. Evidently $\tau^+((b\,e_1 + b^*e_2)^j) = \tau^+(b^je_1 + b^{*j}e_2)$ identifies with $\tau(b^j)$ for all $j \geq 0$, so that $k\tau^+$ satisfies the (k+1)-th symmetric identity on all $*$-fixed elements of $\mathcal{D}$ if and only if $k\tau$ satisfies the identity on <u>all</u> elements of $\mathcal{D}$. Now $\mathcal{D}$ is central simple over Z, and we have seen in §III.4 that $k\tau$ satisfies the (k+1)-symmetric identity on $\mathcal{D}$ if and only if $d|k$. Since $k \leq d$ in our setting, we see that there <u>is no $k\lambda_n$-admissible L_0-module</u> if $k < d$. For $k = d$, it follows from the above and from §III.4 <u>that</u> $d\tau^+$ satisfies the (d+1)-th symmetric identity on $*$-fixed elements of $\mathcal{D}$, and indeed on the $*$-fixed elements for <u>every</u> involution $*$ of second kind (then necessarily $\zeta^* = -\zeta$). Now it follows as before that $d\tau^+$ satisfies the (d+1)-th symmetric identity on every product ab of $*$-fixed elements a,b: Namely $c \to b^{-1}c^*b$ is also an involution of second kind, fixing ab if $a^* = a$, $b^* = b \neq 0$ in $\mathcal{D}$. Thus there exists a unique <u>$d\lambda_n$-admissible irreducible L_0-module</u> (for $n > 1$); <u>it is one-dimensional over</u> F, <u>and</u> $g \in L_0$ <u>acts by</u> $d\tau^+(a_1 +\ldots+ a_n)$.

Now let $1 < j < n$, and consider $k\lambda_j$-admissible L_0-modules $\mathcal{W}$. Here $\mathcal{W}$ must be annihilated by all $[a\,E_{n+1,n}, b\,E_{n,n+1}] = ab\,E_{n+1,n+1} - ba\,E_{nn}$, where a,b are $*$-fixed. In particular, $\mathcal{W}$ is annihilated by all elements $a\,E_{n+1,n+1} - a\,E_{nn}$, $a^* = a$, and by all $[ab]\,E_{n+1,n+1} + [ab]\,E_{nn}$, where a and b are $*$-fixed. If $c_1, c_2 \in \mathcal{D}$, $c_1 = a_1 + b_1\zeta, c_2 = a_2 + b_2\zeta$, where $a_i{}^* = a_i, b_i{}^* = b_i$, we have $[c_1c_2] = [a_1,b_2]\zeta + [b_1,a_2]\zeta + [a_1,a_2] + \gamma[b_1,b_2]$. The first two terms are in $\mathcal{A}$, the second two in $[\mathcal{A}\mathcal{A}]$. Thus $\mathcal{A} + [\mathcal{D}\mathcal{D}] = \mathcal{A} + [\mathcal{A}\mathcal{A}]$, and $\mathcal{W}$ is annihilated by all $c\,E_{n+1,n+1} - c^*E_{n,n}$ for $c \in \mathcal{A} + [\mathcal{D}\mathcal{D}]$. The action on $\mathcal{W}$ of the general element g of L_0 coincides with that of

$$\sum_{i=1}^{n-1}(-a_i{}^*E_{ii} + a_iE_{2n+1-i,2n+1-i}) + \tau^-(a_n)(\zeta E_{nn} + \zeta E_{n+1,n+1}).$$

As in §V.2, continuation of the above analysis to $\alpha_{n-1},\ldots,\alpha_{j+1}$ and application of earlier procedures to $\alpha_1, \ldots, \alpha_{j-1}$ yields that the action of g on $\mathcal{W}$ agrees with that of

$$\sum_{i=j+1}^{n} \tau^-(a_i)(\zeta E_{j+1,j+1} + \zeta E_{2n-j,2n-j})$$

$$+ \sum_{i=1}^{j} \tau^+(a_i)(-E_{jj} + E_{2n+1-j,2n+1-j})$$

$$+ \sum_{i=1}^{j} \tau^-(a_i)(\zeta E_{jj} + \zeta E_{2n+1-j,2n+1-j})$$

$$= \sum_{i=1}^{j} \tau^+(a_i)(-E_{jj} + E_{2n+1-j,2n+1-j})$$

$$- \sum_{i=1}^{j} \tau^-(a_i)(-\zeta E_{jj} + \zeta E_{j+1,j+1} + \zeta E_{2n-j,2n-j} - \zeta E_{2n+1-j,2n+1-j}) .$$

In particular, the action of H_j agrees with that of $-E_{jj} + E_{2n+1-j,2n+1-j}$, and is multiplication by k. Thus the action of the first sum in this last expression is $k \sum_{i=1}^{j} \tau^+(a_i)$.

For $a \in \mathcal{D}$, let $\rho(a)$ be the action on the $k\lambda j$-admissible L_0-module $\mathcal{W}$ of

$$- a^* E_{jj} + a E_{2n+1-j,2n+1-j} - \tau^-(a)(\zeta E_{j+1,j+1} + \zeta E_{2n-j,2n-j}),$$

an element of L_0. Then ρ is F-linear: $\mathcal{D} \to \mathrm{End}_F(\mathcal{W})$, $\rho(1) = kI$, and from the above

$$\rho(a) = k\tau^+(a)I + \tau^-(a)\ \rho(\zeta),$$

so $\rho([ab]) = 0 = [\rho(a), \rho(b)]$. Assuming $\mathcal{W}$ irreducible it follows that the enveloping associative algebra of $\rho(\mathcal{D})$ in $\mathrm{End}_F(\mathcal{W})$ is a (commutative) field.

The mapping ρ satisfies the (k+1)-th symmetric identity. This is seen as in previous cases, by noting that for $a \in \mathcal{D}$, $\psi((-a^* E_{j+1,j} + a E_{2n+1-j,2n-j})(-E_{j,j+1} + E_{2n-j,2n+1-j}))$ acts on $\mathcal{W}$ by $\rho(a)$, and that for commuting $a_1, \ldots, a_t \in \mathcal{D}$, letting $f_t(a_1, \ldots, a_t) \in \mathrm{End}_F(\mathcal{W})$ be the action of

$$\psi((-a_1^* E_{j+1,j} + a_1 E_{2n+1-j,2n-j}) \cdots (-a_t^* E_{j+1,j} + a_t E_{2n+1-j,2n-j}) \cdot$$
$$(-E_{j,j+1} + E_{2n-j,2n+2-j})^t) \in \mathcal{U}(L_0) ,$$

these f_t satisfy our fundamental recursion (1) of Chapter III with $f_1 = \rho$. The necessary condition for admissibility, that $f_{k+1}(a, \ldots, a) = 0$, yields, by (6) of Chapter III, that ρ satisfies the (k+1)-th symmetric identity. By §V.3 it follows that $\mathcal{W}$ is an irreducible $S^k(\mathcal{D})$-module (where it should be emphasized that the ground field is F, not Z).

The enveloping algebra of $\rho(\mathcal{D})$ in $\mathrm{End}_F(\mathcal{W})$ is accordingly a commutative minimal ideal in $S^k(\mathcal{D})$. From §IV.4 it follows that $k = d$ and $\rho(\zeta)^2 = d^2\gamma$, so that setting $\varphi(a) = d^{-1}\rho(a)$ for $a \in \mathcal{D}$, we have an isomorphism $\varphi: Z \to \mathrm{End}_F(\mathcal{W})$ ($\varphi([\mathcal{D}\mathcal{D}]) = 0$). Thus $\mathcal{W}$ is a one-dimensional vector space over Z, and the action of g on $\mathcal{W}$ is scalar multiplication by $d\tau(a_1 + \ldots + a_j)$. In particular, there are no $k\lambda_j$-admissible irreducible L_0-modules for $k < d$, and that (if one exists) for $k = d$ is $\mathcal{W}$ as described above.

To see that our $\mathcal{W}$ is indeed $d\lambda_j$-admissible is done as in previous cases; the only step that is not absolutely routine by now is to show that for $a_1, \ldots, a_t, b$ in $\mathcal{D}$ such that $a_1b, \ldots, a_tb$ commute, the action on the $\mathcal{U}(L_0)$-module $\mathcal{W}$ of

$$\psi((-a_1^* E_{j+1,j} + a_1 E_{2n+1-j,2n-j}) \ldots (-a_t^* E_{j+1,j} + a_t E_{2n+1-j,2n-j}) \cdot$$
$$(-b^* E_{j,j+1} + b\, E_{2n-j,2n+1-j})^t)$$ is $f_t(a_1b, \ldots, a_tb)$, where f_t is obtained by the fundamental recursion III.(1) from $f_1 = \rho = d\tau$. The (d+1)-th symmetric identity for $d\tau$ then yields that $\mathcal{W}$ is annihilated by all

$$\psi((e_{-\alpha_j})^{d+1}(e_{\alpha_j})^{d+1}) ,$$

the only non-trivial one of our conditions for admissibility.

3. CONSTRUCTION OF REPRESENTATIONS: WEIGHTS $k\lambda_1$.

We assume $n > 1$. Then the effect of g on an admissible L_0-module for $\lambda = k\lambda_1$ is equal to that of

$$-a_1^* E_{11} + a_1 E_{2n,2n} + \sum_{j=2}^{n} \tau^-(a_j)(\zeta E_{22} + \zeta E_{2n-1,2n-1})$$
$$= -a_1^* E_{11} + a_1 E_{2n,2n} - \tau^-(a_1)(\zeta E_{22} + \zeta E_{2n-1,2n-1}) .$$

Denote by $\rho(a_1)$ the corresponding F-endomorphism of $\mathcal{W}$; then $\rho(1) = kI$, $\rho([ab]) = [\rho(a), \rho(b)]$, and the admissibility yields exactly as before that ρ satisfies the (k+1)-th symmetric identity. Thus $\mathcal{W}$ is an $S^k(\mathcal{D})$-module. Also, the previous reasoning shows that if $\mathcal{W}$ is any irreducible $S^k(\mathcal{D})$-module, then $\mathcal{W}$ becomes a $k\lambda_1$-admissible irreducible L_0-module by defining the effect on $\mathcal{W}$ of $g \in L_0$ to be that of $\rho_k(a_1) \in S^k(\mathcal{D})$. The generation of $S^k(\mathcal{D})$ by $\rho_k(\mathcal{D})$ shows that $\mathcal{W}$ and $\mathcal{W}'$ are isomorphic $S^k(\mathcal{D})$-modules if and only if they are isomorphic L_0-modules.

For $k \leq d$, the structure of $S^k(\mathcal{D})$ and that of its modules have been studied in §IV.4. In an irreducible $S^k(\mathcal{D})$-module $\mathcal{W}$, $\rho_k(\zeta)^2 = (k-2r)^2\gamma I$, where r is an integer $\le [\frac{k}{2}]$, and if $r = k/2$, $\rho_k(\zeta) = 0$. Upon a splitting extension of the base field, to K, the ideal in $S^k(\mathcal{D})_K$ complementary to the kernel of representations for which r is such that this relation holds becomes

$$(S^r(\mathcal{D}_K e_1) \otimes_K S^{k-r}(\mathcal{D}_K e_2)) \oplus (S^{k-r}(\mathcal{D}_K e_1) \otimes S^r(\mathcal{D}_K e_2)) ,$$

only one term being present if $k = 2r$. From IV.3 we see that $S^k(\mathcal{D})_K = S^k(\mathcal{D}_K)$ is a sum of $\Sigma_{r=0}^{k} p(r)p(k-r)$ minimal ideals, each a split central simple algebra. Upon splitting of $\mathcal{D}$ by K, the action of $(L_0)_K$ on $\mathcal{W}_K$ is obtained from a sum of minimal right ideals in $S^k(\mathcal{D}_K)$. We shall use this information more decisively in the case $n = 1$.

Now let $n = 1$. When $d = 1$, our Lie algebra L is the split three-dimensional algebra, and need not be considered further. For $1 \le k \le d$, the irreducible $S^k(\mathcal{D})$-modules are seen as above to afford a family of irreducible $k\lambda_1$-admissible L_0-modules. The extensions of these to the splitting field K account for $\Sigma_{r=0}^{k} p(r)p(k-r)$ irreducible L_K-modules, whose highest weights all have $k\lambda_1$ as restriction to T. Questions of equivalence are considered below.

In this case the fundamental dominant weights for L_K are

$$\pi_1, \ldots, \pi_{2d-1}, \quad \text{with} \quad \Sigma_{j=1}^{d-1} m_j \pi_j\big|_T = (\Sigma j m_j)\lambda_1 =$$

$$(\Sigma_{j=1}^{d-1} m_j \pi_{2d-j})\big|_T, \quad \text{with} \quad \pi_d\big|_T = d\lambda_1.$$ Thus for $k < d$, the highest weights that restrict to $k\lambda_1$ are those of the form $\Sigma_{j=1}^{d-1} m_j\pi_j + \Sigma_{j=1}^{d-1} n_j \pi_{2d-j}$, where $\Sigma_{j=1}^{d-1} j(m_j + n_j) = k$; for $k = d$, one has all these, together with π_d. For $k < d$, it is clear that the number of such dominant integral functions is again $\Sigma_{r=0}^{k} p(r)p(k-r)$, so that Proposition II.2 yields that $S^k(\mathcal{D})$ contains all irreducible $k\lambda_1$-admissible L_0-modules. For $k = d$, those weights of L_K other than π_d, having $d\lambda_1$ as restriction to T, correspond to partitions $d = r + (d-r)$, $0 \le r \le d$, and for each of these with $0 < r < d$, to a combination of an arbitrary partition of r with one of $d-r$; for $r = d$, to a partition of d other than the extreme partition $0\cdot 1 + 0\cdot 2 + \ldots + 1\cdot d$, and likewise for $r = 0$. The total set of such weights consists of $\Sigma_{r=1}^{d-1} p(r)p(d-r) + 2(p(d) - 1)$ weights of L_K. When π_d is adjoined we see that the totality of weights having $d\lambda_1$ as restriction is $\Sigma_{r=0}^{d} p(r)p(d-r) - 1$. We shall be able to apply Proposition II.2 again if we show that there is <u>exactly</u> <u>one</u> <u>duplication</u> among the irreducible L_K-modules resulting from the $(L_0)_K$-modules afforded by $S^d(\mathcal{D})_K$. In the notation of §IV.3, the latter are in the $d+1$ ideals $S^d(\mathcal{D}_K)e_{r,s}$,

with $r + s = d$, each of these being isomorphic to $S^r(\mathcal{D}_K e_1) \otimes_K S^s(\mathcal{D}_K e_2)$.

Let S_0 be the K-subalgebra of $S^d(\mathcal{D}_K)$ generated by $\rho_d(F1 + [\mathcal{D}\mathcal{D}])$, and let $\varphi_1, \ldots, \varphi_t$ be the projections of $S^d(\mathcal{D}_K)$ onto its minimal two-sided ideals. Now ρ_d is an isomorphism of F-Lie algebras from $\mathcal{D}$ onto its image in $S^k(\mathcal{D})$. Because $d > 1$, $[\mathcal{D}\mathcal{D}]$ is not commutative, and in fact $[\mathcal{D}\mathcal{D}]_K \cong [\mathcal{D}_K e_1, \mathcal{D}_K e_1] \oplus [\mathcal{D}_K e_2, \mathcal{D}_K e_2]$ as K-Lie algbera, a sum of two simple ideals. Thus $\varphi_i(\rho_d(F1 + [\mathcal{D}\mathcal{D}]))$ is a non-commutative set unless $\varphi_i(\rho_d([\mathcal{D}\mathcal{D}])) = 0$. If the image of φ_i lies in $S^d(\mathcal{D}_K)e_{r,s}$ then $\varphi_i(\rho_d(\mathcal{D}))$ generates this image as K-algebra, and $\varphi_i(\rho_d(\zeta)) = \lambda(r-s)\,1$ in $\varphi_i(S^d(\mathcal{D}_K))$, where $\lambda \in K$ has $\lambda^2 = \zeta^2$, as in §IV.3. Thus the image of φ_i is already equal to $\varphi_i(S_0)$, and is generated as K-algebra by $\varphi_i(\rho_d(F1 + [\mathcal{D}\mathcal{D}]))$. The image $\varphi_i(\rho_d([\mathcal{D}\mathcal{D}]))$ is therefore either zero or a K-irreducible set of endomorphisms of a faithful irreducible module for $\varphi_i(S^d(\mathcal{D}_K))$. All those φ_i of the latter type afford inequivalent irreducible $(L_0)_K$-modules, none of which can be isomorphic to one of the former type.

The former φ_i have $\varphi_i(S^d(\mathcal{D}_K)) = \varphi_i(\rho_d(K1 + K\zeta))$ commutative. By §IV.4, there are exactly two such φ_i, one with image in $S^d(\mathcal{D}_K)e_{d,0}$ and the other in $S^d(\mathcal{D}_K)e_{0,d}$. Here one has $\varphi_i(\rho_d(\xi 1 + \eta\xi + c_0)) = \xi d1 \pm \eta\lambda d1$, respectively, where $c_0 \in [\mathcal{D}\mathcal{D}]$. It is clear that the corresponding irreducible $(L_0)_K$-modules, each of K-dimension one, are isomorphic, even though the $S^d(\mathcal{D}_K)$-modules are not. Now it has been shown that $S^d(\mathcal{D}_K)$ affords $\Sigma_{r=0}^d p(r)p(d-r) - 1$ inequivalent irreducible $(L_0)_K$-modules, on each of which the highest weight has restriction $d\lambda_1$ to T. Proposition II.2 applies to enable us to reach the same conclusion as for $n > 1$:

For $1 \le k \le d$, <u>the irreducible</u> $k\lambda_1$-<u>admissible</u> L_0-<u>modules are obtained by letting the general element</u> g <u>of</u> L_0 <u>act on an irreducible</u> $S^k(\mathcal{D})$-<u>module</u> W <u>by</u> $\rho_k(a_1) \in S^k(\mathcal{D})$.

4. CONSTRUCTION OF REPRESENTATIONS. WEIGHTS $k\lambda_j + (d-k)\lambda_{j+1}$.

Here $1 \le k \le d$, $1 \le j \le n-1$. For $1 < j < n-1$, the action on an irreducible λ-admissible L_0-module of our general $g \in L_0$ agrees with that of $-\tau(a_1 + \ldots + a_j)^* E_{jj} + \tau(a_1 + \ldots + a_j)\, E_{2n+1-j,2n+1-j} + \Sigma_{i=j+1}^n (-a_i^* E_{ii} + a_i E_{2n+1-i,2n+1-i})$, and further with that of

$$- \tau(a_1 + \ldots + a_j)^* E_{jj} + \tau(a_1 + \ldots + a_j)\, E_{2n+1-j,2n+1-j}$$
$$- a_{j+1}^* E_{j+1,j+1} + a_{j+1} E_{2n-j,2n-j}$$
$$+ \tau^-(a_{j+1} + \ldots + a_n)(\zeta E_{j+2,j+2} + \zeta E_{2n-1-j,2n-1-j}) .$$

For $a \in \mathcal{D}$ and $\mathcal{W}$ an irreducible λ-admissible L_0-module, we let $\rho(a) \in \mathrm{End}_F(\mathcal{W})$ be the action on $\mathcal{W}$ of

$$-a^* E_{j+1,j+1} + a\, E_{2n-j,2n-j} - \tau^-(a)(\zeta E_{j+2,j+2} + \zeta E_{2n-1-j,2n-1-j}) ,$$

an element of L_0. Then $\rho([ab]) = [\rho(a), \rho(b)]$; $\rho(1)$ is the action of $-E_{j+1,j+1} + E_{2n-j,2n-j}$, which by the above agrees with that of H_{j+1} so is $(d-k)I$. Finally, it follows as before that ρ satisfies the $(d-k+1)$-th symmetric identity.

Now let $A = \mathcal{D}(Z)$, over Z, in Proposition IV.2 replacing k by $r = d-k$, and obtaining a central simple algebra $\mathcal{B}$ over Z, $(\mathcal{B}:Z) = \binom{d}{k}^2$, and a Z-linear mapping $\rho:\mathcal{D}(Z) \to \mathcal{B}$ which is a Lie morphism with $\rho(1) = (d-k)1$, satisfying the $(d-k+1)$-th symmetric identity, and such that $d\tau - \rho$ satisfies the $(k+1)$-th symmetric identity. We let $\mathcal{W}$ be an irreducible right $\mathcal{B}$-module, identifying $\mathcal{B}$ with an irreducible Z-algebra of Z-endomorphisms of $\mathcal{W}$. For $w \in \mathcal{W}$, g our general element of L_0, define

$$w \cdot g = w\, \rho(a_{j+1}) - d\, \tau(\Sigma_{i=1}^{j}\, a_i)w .$$

Then $\mathcal{W}$ is evidently an irreducible L_0-module, the action of L_0 being centralized by Z. Moreover, $wH_j = -w\,\rho(1) + d\tau(1)w = kw$, and $wH_{j+1} = w\,\rho(1) = (d-k)w$, while $w \cdot H_i = 0$ otherwise. That is, $\mathcal{W}$ has weight λ.

An argument identical with those previously given for this purpose shows that $\mathcal{W}$ is λ-admissible. We now claim that $\mathcal{W}$ is the unique λ-admissible irreducible L_0-module. From the list of weights in the split case and from Proposition II.2 we see first that there are only two irreducible L_K-modules whose highest weights have λ as restriction to T, and that it therefore suffices to show that $\mathcal{W}_K$ has inequivalent irreducible $(L_0)_K$-submodules. The action of $\zeta^* E_{jj} - \zeta^* E_{j+1,j+1} + \zeta E_{2n-j,2n-j} - \zeta E_{2n+1-j,2n+2-j} \in L_0$ on $\mathcal{W}$ is $\rho(\zeta) - d\tau(\zeta) = -k\zeta$, so we have an element $x \in L_0$ with $wx = \zeta w$ for all $w \in \mathcal{W}$. With K an extension and $\zeta^2 = \lambda^2$, $\lambda \in K$, we have, in $K \otimes_F \mathcal{W}$, that $w_1 = 1 \otimes \zeta w - \lambda \otimes w$ is sent to $-\lambda w_1$ by x, while $w_2 = 1 \otimes \zeta w + \lambda \otimes w$ is sent to λw_2. Now the subspace $\mathcal{W}_1$ of $K \otimes \mathcal{W}$ annihilated by $x + \lambda$ and that, $\mathcal{W}_2$, annihilated by $x - \lambda$ are $(L_C)_K$-submodules because x is central in L_0, and clearly have no isomorphic submodules. Thus $\mathcal{W}_K$ has (two) inequivalent irreducible $(L_0)_K$-submodules, and our underlined assertion follows.

The remaining cases require only slight modifications of the above. For $\lambda = k\lambda_{n-1} + (d-k)\lambda_n$, $n > 2$, one simply chooses $\mathcal{W}$ to be the irreducible $S^{d-k}(\mathcal{D})$-module above, with $w \cdot g = w\,\rho(a_n) + d\,\tau(\,\Sigma_{i=1}^{n-1}\, a_i)w$, exactly as above.

For $\lambda = k\lambda_1 + (d-k)\lambda_2$, we define the action of g on W to be $\rho(a_1 + a_2) + \sigma(a_1) = \rho(a_2) + d\tau(a_1)$, again obtaining the unique λ-admissible L_0-module. The unified construction of this section reads as follows:

PROPOSITION VI.1. *For* $n > 1$, $1 \le i < n$, $1 \le k < d$, *there is a unique irreducible* $k\lambda_i + (d-k)\lambda_{i+1}$*-admissible* L_0*-module* W. W *is realized as a minimal right ideal in that minimal two-sided ideal* B *of* $S^{d-k}(\mathcal{D}(Z))$ *such that if* $\rho(a)$ *is the projection on* B *of* $\rho_{d-k}(a)$, $a \in \mathcal{D}(Z)$, *then* $d\tau - \rho$ *satisfies the* $(k+1)$-th *symmetric identity, as in Proposition* IV.1.

5. SUMMARY.

With $\mathcal{D}$, Z, L, L_0, T, $\{\alpha_i\}$ as above, we can now carry out a further reduction in our fundamental set. The language is that of §V.7:

THEOREM VI.1. *For* $n \ge 1$, *a fundamental set for* L *consists of the set of weights* $\{\lambda_1, d\lambda_j (1 < j \le n), k\lambda_j + (d-k)\lambda_{j+1} (1 \le j \le n-1, 1 \le k < d)\}$ *and, for each weight* μ *in the set, of a single* μ*-admissible irreducible* L_0*-module* $W^{(\mu)}$. *The underlying space of* $W^{(\mu)}$ *is as follows:*

$\mu = \lambda_1 : W^{(\mu)} = \mathcal{D}$

$\mu = d\lambda_j,\ 1 < j < n : W^{(\mu)} = Z.$

$\mu = d\lambda_n : W^{(\mu)} = F.$

$\mu = k\lambda_j + (d-k)\lambda_{j+1} : W^{(\mu)}$ is a certain minimal right ideal in $S^{d-k}(\mathcal{D}(Z))$, as described in Proposition VI.1.

The action of $g \in L_0$ *on* $W^{(\mu)}$ *is given in* §§2-4.

The only new matter here is the reduction from all $k\lambda_1$, $1 \le k \le d$, to λ_1. We have seen in §3 that minimal right ideals in $S^k(\mathcal{D})$ afford all $k\lambda_1$-admissible irreducible L_0-modules, the action of $g \in L_0$ being right multiplication by $\rho_k(a_1)$. As in previous cases, since $S^k(\mathcal{D})$ is an $S^k(\mathcal{D})$-submodule of $\otimes^k \mathcal{D} = \otimes^k S^1(\mathcal{D})$, $S^1(\mathcal{D})$ being the unique λ_1-admissible irreducible module, and since the k-fold tensor power of $\mathcal{D}$ has g acting by $\rho_k(a_1) \in S^k(\mathcal{D})$, all $k\lambda_1$-admissible irreducible modules occur as submodules of $\otimes^k \mathcal{D}$. This completes the proof.

CHAPTER VII

MODULES FOR LIE ALGEBRAS OF QUADRATIC FORMS

1. THE LIE ALGEBRAS; FUNDAMENTAL WEIGHTS.

The remaining "great class" of central simple Lie algebras with reduced root system is that of the Lie algebras of skew transformations with respect to a nondegenerate quadratic form of positive (but not maximal) index. That is, V is an F-vector space of finite dimension and (u,v) is a nondegenerate symmetric bilinear form on V of Witt index $n > 0$, with $[V:F] > 2n+1$. The Lie algebra L is the set of F-endomorphisms T of V with $(uT, v) + (u, vT) = 0$ for $u,v \in V$. Such are the Lie algebras of relative type B_n of §V.3 of [26] for $n \geq 3$. For $n = 2$, they are those of §V.3c of [26] for which the Jordan algebra "A" is F itself, and for n=1 those of §V.3.d for which "A" is the Jordan algebra of a quadratic form in $q > 1$ variables, not representing 1.

The case $[V:F] = 2n+1$ is the split case of type B_n; we let $[V:F] = 2n+1+q$, where $q > 0$. Let $e_1,\ldots, e_n$ be a basis for a maximal totally isotropic subspace of V, and let $e_{2n+1+q},\ldots, e_{n+q+2}$ (in that order) be a dual basis for a dual totally isotropic subspace. Let e_{n+1} be a non-zero vector orthogonal to these two subspaces; then necessarily e_{n+1} is non-isotropic, so that we may multiply the form by a scalar without changing L assume $(e_{n+1}, e_{n+1}) = 1$. (The vectors already chosen are easily adapted to this change.) Let B be the subspace of V orthogonal to all $2n+1$ vectors chosen so far. Then B is anisotropic of dimension q; in fact, $B + Fe_{n+1}$ is anisotropic, so $(b,b) \neq -1$ for all $b \in B$.

Let T be the subalgebra of L annihilating B and acting diagonally on $B^{\perp}$ with respect to the basis of $2n+1$ vectors chosen above. Then T is commutative with basis the combinations $T_i = E_{2n+2+q-i,2n+2+q-i} - E_{ii}$ of matrix units, $1 \leq i \leq n$. This T is a "maximal split torus" in L; its centralizer L_0 is identified (see [26], p. 186) with

$$K(B) + T + e_{n+1} \otimes B$$

where $K(B)$ is the set of elements of L stabilizing B and annihilating $B^{\perp}$, and $v \otimes b$, for $b \in B$, $v \in B^{\perp}$, is the element of L sending $u + c$ $(u \in B^{\perp}, c \in B)$ to $(u,v)b - (c,b)v$.

A fundamental system of roots for L relative to T is obtained by letting L_{α_i}, $1 \leq i < n$, be the one-dimensional space $F(E_{i,i+1} - E_{2n+q+1-i,2n+q+2-i})$, and L_{α_n} the space $F(E_{n,n+1} - E_{n+1,n+q+2}) + e_{n+q+2} \otimes B$, of dimension $q + 1$. The corresponding $L_{-\alpha_i}$, $L_{-\alpha_n}$ are $F(E_{i+1,i} - E_{2n+q+2-i,2n+q+1-i})$ and $F(E_{n+1,n} - E_{n+q+2,n+1}) + e_n \otimes B$. The uniquely determined $H_j \in T$ with $H_j = [X_jY_j]$, $X_j \in L_{\alpha_j}$, $Y_j \in L_{-\alpha_j}$ a[illegible]) = 2, are $H_i = T_{i+1} - T_i$, $1 \leq i < n$, and $H_n = 2T_n$.

According as q is odd or even, splitting of L results in a split Lie algebra of type D or B, one that is simple except when $n = 1$, $q = 1$. Such an explicit splitting is given in Appendix 4, where a fundamental system $\gamma_1, \ldots, \gamma_\ell$ of roots relative to a Cartan subalgebra H containing T is given. If $q = 2r - 1$, $\ell = n + r$, and we have the diagram

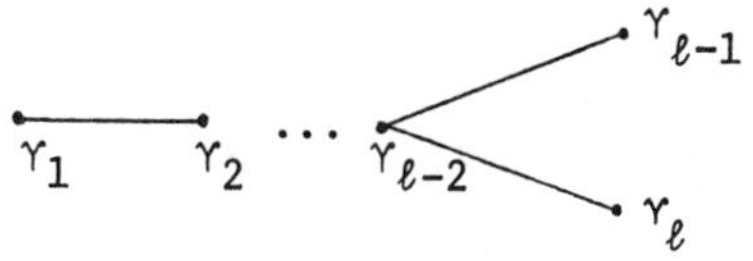

with $\gamma_i|_T = \alpha_i$, $1 \leq i \leq n$ and $\gamma_i|_T = 0$ if $i > n$. The corresponding fundamental weights are

$$\pi_i = \sum_{j=1}^{i} j\gamma_j + i \sum_{j=i+1}^{\ell-2} \gamma_j + \frac{i}{2}(\gamma_{\ell-1} + \gamma_\ell), \quad 1 \leq i \leq \ell - 2;$$

$$\pi_{\ell-1} = \frac{1}{2} \sum_{j=1}^{\ell-2} j\gamma_j + \frac{1}{4}\ell\gamma_{\ell-1} + \frac{1}{4}(\ell-2)\gamma_\ell \;; \pi_\ell = \pi_{\ell-1} + \frac{1}{2}(\gamma_\ell - \gamma_{\ell-1}) .$$

With $\lambda_i(H_j) = \delta_{ij}$, we have $\pi_i|_T = \lambda_i$ for $1 \leq i < n$, $\pi_{n+j}|_T = 2\lambda_n$, $0 \leq j \leq r-2$, and $\pi_{\ell-1}|_T = \pi_\ell|_T = \lambda_n$, provided $r > 1$. If $r = 1$, $\pi_{\ell-2}|_T = \pi_{n-1}|_T = \lambda_{n-1}$, and again $\pi_{\ell-1}|_T = \pi_\ell|_T = \lambda_n$. Hence the restrictions to T of the fundamental π_i are $\{\lambda_1, \ldots, \lambda_n, 2\lambda_n\}$ if $r > 1$, and $\{\lambda_1, \ldots, \lambda_n\}$ if $r = 1$.

If $q = 2r$, we again have $\ell = n + r$, with diagram

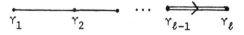

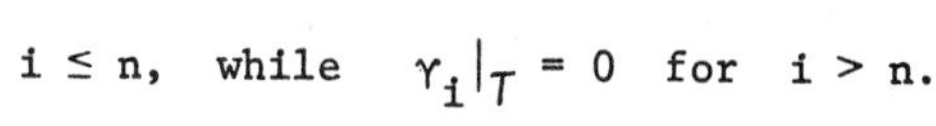

Again $\gamma_i|_T = \alpha_i$, $1 \leq i \leq n$, while $\gamma_i|_T = 0$ for $i > n$. The fundamental weights are

$$\pi_i = \sum_{j=1}^{i} j\gamma_j + i \sum_{j=i+1}^{\ell} \gamma_j, \quad i < \ell, \quad \text{and}$$

$$\pi_\ell = \frac{1}{2} \sum_{j=1}^{\ell} j\gamma_j \quad .$$

Thus $\pi_i|_T = \lambda_i$, $i < n$; $\pi_{n+j}|_T = 2\lambda_n$, $0 \le j < r$; $\pi_\ell|_T = \lambda_n$. Again the weights we must investigate are $\{\lambda_1, \ldots, \lambda_n, 2\lambda_n\}$.

2. REPRESENTATIONS WITH HIGHEST WEIGHT λ_i, $i < n$.

If W is the highest weight space for such a representation, we must have $w H_j = \delta_{ij} w$ for $w \in W$ and all j. The admissibility conditions force $w e_{-\alpha_j} e_{\alpha_j} = 0$ for all $e_{\pm\alpha_j} \in L_{\pm\alpha_j}$, $j \ne i$, while $w e_{-\alpha_i}^2 e_{\alpha_i}^2 = 0$. For $j < n$, $j \ne i$, $[L_{-\alpha_j} L_{\alpha_j}] = F H_j$ ($L_{\pm\alpha_j}$ has dimension one), and the criterion for admissibility is redundant since $W H_j = 0$.

For $j = n$, admissibility requires that

$$0 = w(\mu(E_{n+1,n} - E_{n+q+2,n+1}) + e_n \otimes b)(\nu(E_{n,n+1} - E_{n+1,n+q+2}) + e_{n+q+2} \otimes c)$$

for all μ, $\nu \in F$; $w \in W$; $b, c \in B$. From choosing $\mu = 1$, $b = 0$, $\nu = 0$, and the fact that $[E_{n+1,n} - E_{n+q+2,n+1}, e_{n+q+2} \otimes c] = e_{n+1} \otimes c$, we have $w(e_{n+1} \otimes c) = 0$ for all w, all c. Then setting $\mu = \nu = 0$, we have $[e_n \otimes b, e_{n+q+2} \otimes c] = S_{b,c} + (b,c)^*(E_{n+q+2,n+q+2} - E_{nn}) = S_{b,c} + (b,c)^* T_n$, $= S_{b,c} - (b,c)T_n$, the form $(b,c)^*$ on B being the negative of our form (b,c) (see §V.3.a of [26] for calculations performed within L in the decomposition used here.) Here $S_{b,c}$ sends $v \in V$ to $(v,c)b - (v,b)c$. Together with the annihilation of W by $H_n = 2T_n$, this gives $wS_{b,c} = 0$ for all $b,c \in B$. These $S_{b,c}$ linearly generate $K(B)$, so $WK(B) = 0$. Thus if $g = S + \Sigma_{j=1}^n \mu_j H_j + e_{n+1} \otimes b$, $S \in K(B)$, $b \in B$, is the general element of L_0, we have for $w \in W$, $wg = \mu_i wH_i = \mu_i w$. Thus <u>the irreducible λ_i-admissible L_0-module must be one-dimensional with this action, so is unique</u>.

To see that a one-dimensional space W with the above action of L_0 is λ_i-admissible is straightforward; Calculations as above show W is annihilated by $[L_{-\alpha_j} L_{\alpha_j}]$ for $j \ne i$, while $w\psi(e_{-\alpha_i}^2 e_{\alpha_i}^2) =$

$$w\psi((\mu(E_{i+1,i} - E_{2n+q+2-i,2n+q+1-i}))^2(\nu(E_{i,i+1} - E_{2n+q+1-i,2n+q+2-i}))^2)$$

$= 2\mu^2\nu^2 wH_i^2 - 2\mu^2\nu^2 wH_i = 0$ since $wH_i = w$. Thus W is λ_i-admissible. The corresponding irreducible L-module is the space of skew-symmetric tensors of rank i over V (or the exterior power $\Lambda^i V$), with highest weight-vector

$$e_{2n+q+2-i} \wedge \cdots \wedge e_{2n+q} \wedge e_{2n+q+1} \quad .$$

3. REPRESENTATIONS WITH HIGHEST WEIGHT λ_n.

For irreducible representations of L with highest weight λ_n and highest weight-space W, we have as in §2 that $0 = wH_i = wL_{-\alpha_i} L_{\alpha_i}$ for all $i < n$, all $w \in W$. From $wH_n = w$ we have as in §2 that

$$w(E_{n+1,n} - E_{n+q+2,n+1})^2(E_{n,n+1} - E_{n+1,n+q+2})^2 = 0$$

for all $w \in W$.

For $b \in B$, let $T_b: W \to W$ send w to $w(e_{n+1} \otimes b)$; then $w[T_b, T_c] = w \cdot S_{b,c}$ by a calculation in L, so that the effect of $K(B)$ on W is determined by the T_b, for $b \in B$. Next consider the necessary condition

$$w(E_{n+1,n} - E_{n+q+2,n+1})^2(e_{n+q+2} \otimes c)^2 = 0 ,$$

which yields

$$2w(T_c)^2 + \frac{(c,c)}{2} wH_n = 0, \quad \text{or}$$

$$4w(T_c)^2 = -(c,c)w = (c,c)^* w$$

Let $P_c = 2T_c$. Then $S_{b,c}$ sends $d \in B$ to $dS_{b,c} = (d,b)^* c - (d,c)^* b$, and $w(P_c)^2 = (c,c)^* w$. By the definition of the Clifford algebra $C\ell(B, *)$ we see that there is a unique homomorphism of $C\ell(B, *)$ into $\mathrm{End}_F(W)$ sending $b \in B$ to P_b. Thus W is a right $C\ell(B, *)$-module, with $wb = wP_b = 2w(e_{n+1} \otimes b)$ for all $w \in W$, $b \in B$. The action of L_0 on W is then determined from this and from $wS_{b,c} = \frac{1}{4} w[P_b, P_c] = \frac{1}{4} w(bc-cb)$, $w(\Sigma \mu_i H_i) = \mu_n w$. Here $bc - cb \in C\ell(B, *)$. It is clear that W is irreducible as L_0-module if and only if W is irreducible as $C\ell(B,*)$-module.

If $q = \dim B$ is <u>even</u>, then $C\ell(B, *)$ is central simple, and W is unique. If q is <u>odd</u>, then either $C\ell(B, *)$ has a split (two-dimensional) center and two inequivalent irreducible modules, or the center of $C\ell(B,*)$ is a quadratic extension field Z of F. In the latter case, the irreducible $C\ell(B, *)$-module is again unique. (For the structure of the Clifford algebras, see [6], [12].)

Now $K(B)$ may be identified with the set of those elements $y = \Sigma_j [b_j c_j]$ in $C\ell(B, *)$, where $b_j, c_j \in B$, the action of y on $d \in B$ sending d to $[dy] = 4 \sum_j dS_{b_j,c_j}$. Thus if we start with an irreducible right $C\ell(B, *)$-module W, and set $w(y + \Sigma \mu_j H_j + e_{n+1} \otimes b) = wy + \mu_n w + \frac{1}{2} wb$, where y is as above, $\mu_j \in F$, $b \in B$, it is clear that W is annihilated by all $\psi(e_{-\alpha_i} e_{\alpha_i})$, $i < n$, and that T acts on W with weight λ_n. To see that W is an L_0-module, note that

$$[y + e_{n+1} \otimes b, y' + e_{n+1} \otimes b'] = [yy']$$
$$+ S_{b,b'} + e_{n+1} \otimes [by'] - e_{n+1} \otimes [b',y]$$

sends $w \in \mathcal{W}$ to

$$w[yy'] + \frac{1}{4} w[bb'] + \frac{1}{2} w[by'] - \frac{1}{2} w[b'y].$$

This is the same as the effect,

$$(wy)y' + \frac{1}{2} (wb)y' + \frac{1}{2} (wy)b' + \frac{1}{4} (wb)b'$$
$$-(wy)y' - \frac{1}{2} (wb')y - \frac{1}{2} (wy')b - \frac{1}{4} (wb')b,$$

of the commutator of the actions on $\mathcal{W}$ of $y + e_{n+1} \otimes b$ and $y' + e_{n+1} \otimes b'$.

The λ_n-admissibility of $\mathcal{W}$ is now only a question of its annihilation by all $\psi(e^2_{-\alpha_n} e^2_{\alpha_n})$. In $\mathcal{U}(L_0)$, this is the projection of an element

$$(\mu(E_{n+1,n} - E_{n+q+2,n+1}) + e_n \otimes b)^2(\nu(E_{n,n+1} - E_{n+1,n+q+2}) + e_{n+q+2} \otimes b')^2$$
$$= \{\mu^2(E_{n+1,n} - E_{n+q+2,n+1})^2 + 2\mu(e_n \otimes b)(E_{n+1,n} - E_{n+q+2,n+1})$$
$$+ (e_n \otimes b)^2\} \ \{\nu^2(E_{n,n+1} - E_{n+1,n+q+2})^2$$
$$+ 2\nu(E_{n,n+1} - E_{n+1,n+q+2})(e_{n+q+2} \otimes b') + (e_{n+q+2} \otimes b')^2\}$$

in $\mathcal{U}(L)$. One verifies directly the annihilation of $\mathcal{W}$ by each of the nine terms in the expansion of the image under ψ of this product. For example,

$$\psi(\mu(e_n \otimes b)(E_{n+1,n} - E_{n+q+2,n+1})(e_{n+q+2} \otimes b')^2)$$
$$= \psi(\mu(e_n \otimes b)(e_{n+q+2} \otimes b')(E_{n+1,n} - E_{n+q+2,n+1})(e_{n+q+2} \otimes b'))$$
$$+ \psi(\mu(e_n \otimes b)(e_{n+1} \otimes b')(e_{n+q+2} \otimes b'))$$
$$= \psi(\mu(S_{b,b'} - (b,b')T_n) \ (e_{n+1} \otimes b'))$$
$$+ \psi(\mu(e_n \otimes b)(e_{n+q+2} \otimes b')(e_{n+1} \otimes b'))$$
$$+ \psi(\mu(e_n \otimes b)[e_{n+1} \otimes b', e_{n+q+2} \otimes b'])$$
$$= \ 2\mu(S_{b,b'} - (b,b')T_n)(e_{n+1} \otimes b')$$
$$+ \mu(b', b') \ ((e_n \otimes b)(E_{n,n+1} - E_{n+1,n+q+2}))$$
$$= \ 2\mu(S_{b,b'} - (b,b')T_n)(e_{n+1} \otimes b')$$
$$+ \ \mu(b',b')e_{n+1} \otimes b.$$

The effect of this element on $w \in \mathcal{B}$ is

$$\frac{\mu}{4} w[bb']b' - \frac{\mu}{2} (b,b')wb' + \frac{\mu}{2} (b', b')wb.$$

In $C\ell(\mathcal{B},*)$, $[bb']b' = bb'^2 - (b'b + bb')b' + bb'^2 = -2(b',b')b + 2(b,b')b'$, from which it follows that the last expression above is zero.

The results of this section are summarized in

PROPOSITION VII.1. *Let* $q = \dim \mathcal{B}$, *and let* $C\ell(\mathcal{B},*)$ *be the Clifford algebra of the negative of the given form on* $\mathcal{B}$. *Then* L *has exactly one irreducible representation of highest weight* λ_n *except when* q *is odd and the center of* $C\ell(\mathcal{B},*)$ *splits. In that case, there are two. The highest weight-space* $\mathcal{W}$ *is an irreducible right* $C\ell(\mathcal{B},*)$*-module, and the action of* L_0 *on* $\mathcal{W}$ *satisfies:*

$$w(e_{n+1}\otimes b) = \frac{1}{2}\,wb;\quad wS_{b,c} = \frac{1}{4}\,w(bc-cb);$$

$$w\Big(\sum_{i=1}^{n}\mu_i H_i\Big) = \mu_n w.$$

The irreducible L-modules in question may be realized as minimal right ideals in the *even* Clifford algebra $C\ell^+(V)$ (see [15], §VII.6 for the split case in a general setting). If q is even, $[V:F]$ is odd, $C\ell^+(V)$ is simple, and our L-module is the irreducible $C\ell^+(V)$-module. If q is odd, $C\ell^+(V)$ is semisimple with a two-dimensional center that splits, or does not, along with that of $C\ell(\mathcal{B},*)$.

4. REPRESENTATIONS OF HIGHEST WEIGHT $2\lambda_n$.

Again defining $wT_b = w(e_{n+1}\otimes b)$ for a $2\lambda_n$-admissible L_0-module $\mathcal{W}$, we again have

$$wS_{b,c} = w[T_b, T_c]\ .$$

From
$$[e_{n+1}\otimes b,\ S_{b,c}] = e_{n+1}\otimes bS_{b,c}$$
$$= (b,b)^*\, e_{n+1}\otimes c - (b,c)^*\, e_{n+1}\otimes b,$$

we have $[T_b[T_b, T_c]] = (b,b)^*\, T_c - (b,c)^*\, T_b$, *i.e.*,

(1) $$T_b^2T_c - 2T_bT_cT_b + T_cT_b^2 = (b,b)^*\, T_c - (b,c)^*\, T_b.$$

Next consider the crucial condition $w\psi(e_{-\alpha_n}^3\, e_{\alpha_n}^3) = 0$ of Chapter I, the notations being those of the last section. We specialize to $\mu = 1$, $\nu = 0$, $b = 0$, thus requiring

(2) $$w\psi((E_{n+1,n} - E_{n+q+2,n+1})^3(e_{n+q+2}\otimes b')^3) = 0.$$

Applying the commutation relations to move one factor $e_{n+q+2}\otimes b'$ to the left, we reduce the left side of (2) to

$$(3)\qquad 3w\psi((e_{n+1}\otimes b')(E_{n+1,n}-E_{n+q+2,n+1})^2(e_{n+q+2}\otimes b')^2)$$
$$-3w\psi((e_n\otimes b')(E_{n+1,n}-E_{n+q+2,n+1})(e_{n+q+2}\otimes b')^2).$$

The first term here is $3wT_{b'}\,\psi((E_{n+1,n}-E_{n+q+2,n+1})^2(e_{n+q+2}\otimes b')^2)$. Now calculations performed as in §3, here with $wH_n = 2w = 2wT_n$ for $w \in \mathcal{W}$, reduce (3) to

$$6wT_{b'}^3 - 6(b', b')^* \, wT_{b'} .$$

Thus we must have

$$(4)\qquad wT_b^3 = (b,b)^* \, wT_b$$

for all $w \in \mathcal{W}$, $b \in \mathcal{B}$. Partial polarization gives

$$(5)\qquad T_bT_cT_b + T_b^2T_c + T_cT_b^2 = (b,b)^*T_c + 2(b,c)^*T_b,$$

and subtracting (1) from (5) yields

$$(6)\qquad T_bT_cT_b = (b,c)^*T_b .$$

The associative algebra generated by the subspace $\mathcal{B}$ with relations $bcb = (b,c)^*b$ has been studied by Duffin, Kemmer [19] and others as a physical formalism, and by Jacobson [17], in the representation theory of Jordan algebras of quadratic forms. It is commonly called a <u>meson algebra</u>; we follow Jacobson in notation, denoting it by $\mathcal{D}(\mathcal{B},*)$. The relation (6) means that the linear map $b \mapsto T_b$ of $\mathcal{B}$ into $\mathrm{End}(\mathcal{W})$ lifts to a unique homomorphism of $\mathcal{D}(\mathcal{B},*)$ into $\mathrm{End}(\mathcal{W})$. Thus $\mathcal{W}$ <u>is a right $\mathcal{D}(\mathcal{B},*)$-module, and we have</u>

$$w(e_{n+1}\otimes b) = wb;\quad wS_{b,c} = w(bc-cb);$$
$$w(\Sigma\,\mu_iH_i) = 2\mu_n w,$$

for all $w \in \mathcal{W}$; $b,c \in \mathcal{B}$; $\mu_i \in F$. It follows that $\mathcal{W}$ is an irreducible $\mathcal{D}(\mathcal{B},*)$-module if and only if $\mathcal{W}$ is an irreducible L_0-module.

One can verify directly that each right $\mathcal{D}(\mathcal{B},*)$-module $\mathcal{W}$ becomes a $2\lambda_n$-admissible L_0-module. We do not display the calculations involved, since the result will follow from a tensor-product reduction developed below. The consequence that should be observed is that the determination of the irreducible $2\lambda_n$-admissible L_0-modules is equivalent to the determination of the irreducible (right) $\mathcal{D}(\mathcal{B},*)$-modules.

The latter have been found by Jacobson ([17], §VII.2). They are all realized as submodules of $\mathcal{D}(\mathcal{B},*)$-modules whose underlying space is $C\ell(\mathcal{B},*)$. First note that if $b \in \mathcal{B}$, the mapping $S_b: x \mapsto xS_b = \frac{1}{2}(bx + xb)$ of $C\ell(\mathcal{B},*)$

into itself satisfies $S_b S_c S_b = (b,c)^* S_b$, so that $b \to S_b$ extends to give $C\ell(\mathcal{B},*)$ a structure of right $\mathcal{D}(\mathcal{B},*)$-module. Jacobson gives a decomposition of this module, amounting to the following:

Fix an orthogonal basis $b_1,\dots, b_q$ for $\mathcal{B}$. For each even integer $2\nu \le q$, let M_ν be the subspace of $C\ell(\mathcal{B},*)$ with basis the products $b_{i_1} b_{i_2} \cdots b_{i_{2\nu}}$, $i_1 < \dots < i_{2\nu}$, and (if $2\nu < q$) the products $b_{i_1} b_{i_2} \cdots b_{i_{2\nu+1}}$, the indices again distinct. Then M_ν is stable under all S_{b_i}, hence is a $\mathcal{D}(\mathcal{B},*)$-submodule. (When $\nu = 0$, $M_0 = F1 + \mathcal{B}$.) If $q = 2r$ is <u>even</u>, these M_ν, $0 \le \nu \le r$, are a complete set of inequivalent irreducible modules for $\mathcal{D}(\mathcal{B},*)$, and all of them are absolutely irreducible. Thus $\mathcal{D}(\mathcal{B},*)$ is a direct product of full matrix algebras over F, of degrees $\binom{2r}{2\nu} + \binom{2r}{2\nu+1} = \binom{2r+1}{2\nu+1}$ for $0 \le \nu \le r$, <u>i.e.</u>, of degrees $\binom{2r+1}{j}$, $0 \le j \le r$ ([17], Theorem 4, p. 269).

If $q = 2r + 1$ is <u>odd</u>, matters are more complicated. For $0 \le \nu \le r$, let M_ν be as above. The central element $z = b_1 \cdots b_q$ of $C\ell(\mathcal{B},*)$ satisfies $b_{i_1} \cdots b_{i_j} z = \pm(b_{i_1}, b_{i_1})^* \cdots (b_{i_j}, b_{i_j})^* b_{k_1} \cdots b_{k_{q-j}}$, where $\{i_1,\dots, i_j\}$ and $\{k_1,\dots, k_{q-j}\}$ partition $\{1,\dots, q\}$. It follows that multiplication by z induces an isomorphism between the $\mathcal{D}(\mathcal{B},*)$-modules M_ν and $M_{r-\nu}$.

One compensates for the resulting duplication of modules by noting that if $S_b' : C\ell(\mathcal{B},*) \to C\ell(\mathcal{B},*)$ is defined for $b \in \mathcal{B}$ by $xS_b' = \frac{1}{2}(xb-bx)$, then again $S_b' S_c' S_b' = (b,c)^* S_b'$; thus $C\ell(\mathcal{B},*)$ carries a second structure of $\mathcal{D}(\mathcal{B},*)$-module, in which $b \in \mathcal{B} \subseteq \mathcal{D}(\mathcal{B},*)$ acts by S_b'. Here we let N_ν, $0 \le \nu \le r+1$, be the subspace of $C\ell(\mathcal{B},*)$ with basis the $b_{i_1} \cdots b_{i_{2\nu-1}}$ and the $b_{i_1} \cdots b_{i_{2\nu}}$ as before (here $N_0 = F\cdot 1$). Then N_ν is seen as before to be a $\mathcal{D}(\mathcal{B},*)$-submodule of $C\ell(\mathcal{B},*)$ in the action of this paragraph, with multiplication by z inducing an isomorphism between N_ν and $N_{r+1-\nu}$.

Each M_ν has dimension $\binom{q+1}{2\nu+1} = \binom{2r+2}{2\nu+1}$, where $0 \le \nu \le [\frac{r}{2}]$ may be assumed, and each N_ν, $0 \le \nu \le [\frac{r+1}{2}]$, has dimension $\binom{2r+2}{2\nu}$. If r is <u>odd</u>, all M_ν are irreducible, indeed absolutely irreducible, as are all N_ν, $\nu < \frac{r+1}{2}$; if r is <u>even</u>, all N_ν and all M_ν, $0 \le \nu < \frac{r}{2}$, are absolutely irreducible. In the former case z is a nontrivial $\mathcal{D}(\mathcal{B},*)$-endomorphism of $N_{\frac{r+1}{2}}$, in the latter of $M_{\frac{r}{2}}$. When the center $F(z)$ of $C\ell(\mathcal{B},*)$ is a field Z, the corresponding $\mathcal{D}(\mathcal{B},*)$-module is irreducible. When $z^2 \in F^2$, on the other hand, this module splits into two absolutely irreducible modules, each of dimension $\frac{1}{2}\binom{2r+2}{r+1}$. All the above afford a complete set of $\mathcal{D}(\mathcal{B},*)$-modules, faithfully representing $\mathcal{D}(\mathcal{B},*)$ on their sum. Thus, if the center Z of $C\ell(\mathcal{B},*)$ splits, $\mathcal{D}(\mathcal{B},*)$ is the direct product of F-matrix algebras of degrees $\binom{2r+2}{j}$, $0 \le j \le r$, and of <u>two</u> F-matrix algebras of degree $\frac{1}{2}\binom{2r+2}{r+1}$. If Z is a quadratic extension field of F, then $\mathcal{D}(\mathcal{B},*)$ is a direct product of F-matrix algebras of degrees $\binom{2r+2}{j}$, $0 \le j \le r$, and of one Z-matrix algebra of degree $\frac{1}{2}\binom{2r+2}{r+1}$ ([17], Theorem 5, p. 272).

Assuming the result, proved in the next section, that these are admissible for $2\lambda_n$, we have

PROPOSITION VII.2: *Let $\mathcal{W}$ be the highest weight-space for an irreducible $\mathcal{L}$-module of highest weight $2\lambda_n$. Then $\mathcal{W}$ carries the structure of irreducible right module for the meson algebra $\mathcal{D}(\mathcal{B},*)$ of the form which is the negative of (b,c) on $\mathcal{B}$, and so is as given above. The action of $\mathcal{L}_0$ on $\mathcal{W}$ in terms of this structure of $\mathcal{D}(\mathcal{B},*)$-module satisfies*

$$w(e_{n+1} \otimes b) = wb; \quad wS_{b,c} = w[b,c];$$

$$w(\Sigma\, \mu_i H_i) = 2\mu_n w.$$

If $q = 2r$ is even, there are $r + 1$ irreducible $\mathcal{L}$-modules of highest weight $2\lambda_n$; if $q = 2r + 1$ and the center Z of $C\ell(\mathcal{B},*)$ does not split, there are $r + 2$; if $q = 2r + 1$ and Z is split, there are $r + 3$ modules.

For $n \le i \le n + t - 1$, where $q + 1 = 2t + 1$ or $2t$, the i-th exterior power of our original vector space V affords one of these representations of $\mathcal{L}$; thus r of them are realized if $q = 2r$ and $r + 1$ if $q = 2r + 1$. The latter representations are absolutely irreducible and become those of highest weights $\pi_n, \pi_{n+1}, \ldots, \pi_{n+t-1}$ on passing to the splitting field K. The remaining modules are not associated with fundamental weights for $\mathcal{L}_K$; for example, when $q = 2r$ the remaining module is absolutely irreducible and becomes that of highest weight $2\pi_{n+r}$ for $\mathcal{L}_K$.

5. SUMMARY. GENERATING MODULES.

To show that the λ-admissible modules of §§2,3 for $\lambda = \lambda_1, \ldots, \lambda_n$ are generators in the sense of § V.7, it suffices to show that all the irreducible $\mathcal{L}_0$-modules of weight $2\lambda_n$ resulting as in §4 from irreducible $\mathcal{D}(\mathcal{B},*)$-modules occur in the decomposition of the tensor product of two irreducible λ_n-admissible modules, as described in Proposition 1. By §II.2, it will follow that the modules of §4 are $2\lambda_n$-admissible.

First let $q = 2r$ and let $\mathcal{W}$ be the irreducible λ_n-admissible

L_0-module of Proposition 1. It suffices to give an injection of $\mathcal{D}(\mathcal{B},*)$ into $\mathrm{End}_F(\mathcal{W} \otimes_F \mathcal{W})$ compatible with the action, defined in §4, of L_0 on $\mathcal{D}(\mathcal{B},*)$-modules and with the tensor product of the action of L_0 on $C\ell(\mathcal{B},*)$-modules, defined in §3. For clarity we replace the first factor $\mathcal{W}$ in the tensor product by an isomorphic $C\ell(\mathcal{B},*)$-module. Thus let M be a minimal left ideal in $C\ell(\mathcal{B},*)$, and let $x \to x^\sigma$ be the involution in $C\ell(\mathcal{B},*)$ that fixes all elements of $\mathcal{B}$. Then M^σ is a minimal right ideal, and becomes an irreducible λ_n-admissible L_0-module as in §3. It follows that M is also an irreducible λ_n-admissible L_0-module, with

$$m(e_{n+1} \otimes b) = \frac{1}{2}\ bm; \quad mS_{b,c} = \frac{1}{4}\ [c,b]m;$$
$$m(\Sigma\mu_i H_i) = \mu_n m.$$

Now $M \otimes_F M^\sigma$ is a $(C\ell(\mathcal{B},*),\ C\ell(\mathcal{B},*))$-bimodule, hence a direct sum of copies of the unique simple $(C\ell(\mathcal{B},*),\ C\ell(\mathcal{B},*))$-bimodule $C\ell(\mathcal{B},*)$. The tensor product action of L_0 on $M \otimes M^\sigma$ is as follows:

$$(m \otimes m')\ (e_{n+1} \otimes b) = \frac{1}{2}\ (bm \otimes m' + m \otimes m'b);$$
$$(m \otimes m')S_{b,c} = \frac{1}{4}\ ([c,b]m \otimes m' + m \otimes m'[b,c]):$$
$$(m \otimes m')(\Sigma \mu_i H_i) = 2\mu_n\ m \otimes m',$$

for $m \in M$, $m' \in M^\sigma$. It will be noted that the action by S_b of §4 on $C\ell(\mathcal{B},*)$, for $b \in \mathcal{B}$, extended to a sum of copies of $C\ell(\mathcal{B},*)$ induces by restriction to elements $m \otimes m'$ of this sum exactly the action of $e_{n+1} \otimes b$ given above. It follows that each irreducible L_0-summand of $M \otimes M^\sigma$ is $2\lambda_n$-admissible. In §4 we saw that the action of $\mathcal{D}(\mathcal{B},*)$ on $C\ell(\mathcal{B},*)$ is faithful and affords all irreducible right $\mathcal{D}(\mathcal{B},*)$-modules; furthermore all irreducible $2\lambda_n$-admissible L_0-modules are irreducible $\mathcal{D}(\mathcal{B},*)$-modules. Consequently, $M \otimes M^\sigma$ contains all irreducible $2\lambda_n$-admissible L_0-modules. For q even, the objective of the first paragraph of this section has been achieved.

As in §4, the situation when $q = 2r + 1$ is more complicated. Write C for $C\ell(\mathcal{B},*)$. In addition to the involution σ fixing $\mathcal{B}$, C admits a second involution τ such that $b^\tau = -b$ for all $b \in \mathcal{B}$. Each of these involutions gives a structure of (right) $C \otimes_F C$-module to C, by

$$u(x \otimes y) = x^\sigma uy \quad \text{resp.} \quad x^\tau uy,$$

for $x,y,u \in C$. We denote these structures by $C^{(\sigma)}$ resp. $C^{(\tau)}$ and use them to analyze $C \otimes_F C$.

With the orthogonal basis $b_1,\ldots, b_q$ for $\mathcal{B}$ as in §4, the central element $z = b_1 \dots b_q \in C$ has $z^\sigma = (-1)^r z$, $z^\tau = (-1)^{r+1} z$. Thus either σ fixes Z while τ does not (if r is even), or vice versa (if r is odd). No generality will be lost by carrying out the analysis from here on for <u>even</u> r, which we assume. Thus $z^2 = \gamma \in F$, and Z splits if and only if $\gamma \in F^2$. The element $e = \frac{1}{2}(1 \otimes 1 + \gamma^{-1} z \otimes z)$ is a central idempotent in $C \otimes C$. For $u \in C^{(\sigma)}$, we have $ue = u$, while $ue = 0$ for $u \in C^{(\tau)}$.

If C is simple, *i.e.*, if Z does not split, $C^{(\sigma)}$ and $C^{(\tau)}$ are irreducible $C \otimes C$-modules; the action of e shows that they are inequivalent. If $A = C \otimes C$, we have $A = eA + (1-e)A$, a direct sum of two-sided ideals, with eA annihilating $C^{(\tau)}$ and $(1-e)A$ annihilating $C^{(\sigma)}$. Evidently the centralizer of the action of A on $C^{(\sigma)}$ or $C^{(\tau)}$ is Z, so that Wedderburn's theorem shows that the image of A in $\mathrm{End}_F(C)$ (for either action) is $\mathrm{End}_Z(C)$, of F-dimension $2[C:Z]^2 = 2\cdot(2^{2r})^2 = 2^{4r+1}$. The kernel (call it $K^{(\sigma)}$, $K^{(\tau)}$ respectively) has dimension $\dim A - 2^{4r+1} = 2^{4r+1}$. But $eA \subseteq K^{(\tau)}$, $(1-e)A \subseteq K^{(\sigma)}$, and $A = eA \oplus (1-e)A$. It follows that $eA = K^{(\tau)}$, $(1-e)A = K^{(\sigma)}$, $A \cong \mathrm{End}_Z(C^{(\sigma)}) \oplus \mathrm{End}_Z(C^{(\tau)})$, and that $C^{(\sigma)}$ and $C^{(\tau)}$ <u>are a complete set of irreducible</u> A-<u>modules</u>.

When $z^2 = \beta^2$ for $\beta \in F$, let $e_1 = \frac{1}{2}(1 + \beta^{-1} z)$, $e_2 = 1 - e_1$, an orthogonal pair of central idempotents in C. If $C_i = e_i C$, for $i = 1,2$, then $C = C_1 \oplus C_2$, a direct sum of two-sided ideals, and $C^{(\sigma)} = C_1^{(\sigma)} \oplus C_2^{(\sigma)}$, $C^{(\tau)} = C_1^{(\tau)} \oplus C_2^{(\tau)}$, Continuing to assume r even, we have that $e_1 \otimes e_1 \in A$ annihilates $C_2^{(\sigma)}$ and $C^{(\tau)}$ and is the identity on $C_1^{(\sigma)}$. Similar observations using $e_1 \otimes e_2$, $e_2 \otimes e_1$, $e_2 \otimes e_2$ show that <u>the four</u> A-<u>modules</u> $C_1^{(\sigma)}$, $C_1^{(\tau)}$, $C_2^{(\sigma)}$, $C_2^{(\tau)}$ <u>are inequivalent</u>.

The structure of Clifford algebras, as in [12], means that C_1 and C_2 are central simple F-algebras of dimension 2^{2r}. Thus $A \cong (C_1 \otimes C_1) \oplus (C_1 \otimes C_2) \oplus (C_2 \otimes C_1) \oplus (C_2 \otimes C_2)$, the summands admitting the irreducible modules $C_1^{(\sigma)}$, $C_2^{(\tau)}$, $C_1^{(\tau)}$, $C_2^{(\sigma)}$, in the same order. <u>These four modules are a complete set of irreducible</u> A-<u>modules</u>.

The mapping $\varphi\colon b \to \frac{1}{2}(1 \otimes b + b \otimes 1)$ from B into A satisfies $\varphi(b)\varphi(c)\varphi(b) = (b,c)^*\varphi(b)$ for all $b, c \in B$, hence extends to a unique homomorphism from the meson algebra $\mathcal{D}(B,*)$ into A. By this mapping, all A-modules become $\mathcal{D}(B,*)$-modules. The citations from [17] given in connection with Proposition 2 show that all irreducible $\mathcal{D}(B,*)$-modules occur as submodules of $C^{(\sigma)}$ or of $C^{(\tau)}$. Now it follows from the preceding paragraphs that <u>all irreducible</u> $\mathcal{D}(B,*)$-<u>modules occur as submodules of irreducible</u> A-<u>modules</u>.

Now let m be a minimal left ideal in C, and n a minimal right ideal. Then $m \otimes n$ has two structures of A-module:

$(m \otimes n)^{(\sigma)}$: $(m \otimes n)(x \otimes y) = x^{\sigma}m \otimes ny$; and

$(m \otimes n)^{(\tau)}$: $(m \otimes n)(x \otimes y) = x^{\tau}m \otimes ny$.

Each of these A-modules is a direct sum of simple A-submodules, isomorphic to one of $C^{(\sigma)}$, $C^{(\tau)}$ if Z is not split, or to one of $C_1^{(\sigma)}$, $C_2^{(\sigma)}$, $C_1^{(\tau)}$, $C_2^{(\tau)}$ otherwise.

The map sending $m \otimes n$ as above to mn is a module-homomorphism into $C^{(\sigma)}$ resp. $C^{(\tau)}$. If C is simple, the image must exhaust C. In this case, $(m \otimes n)^{(\sigma)}$ has $C^{(\sigma)}$ has homomorphic image, hence as submodule; likewise, $(m \otimes n)^{(\tau)}$ contains a submodule isomorphic to $C^{(\tau)}$.

When $C = C_1 \oplus C_2$, we have $m \subseteq C_1$ or $m \subseteq C_2$, and likewise for n. We denote these four cases by $m_1 \otimes n_1$, etc. Then $(m_1 \otimes n_1)^{(\tau)}$ is annihilated by $e_2 \otimes e_1$, $e_1 \otimes e_2$, $e_2 \otimes e_2$, so is a sum of copies of $C_1^{(\sigma)}$, and by the same reasoning $(m_1 \otimes n_1)^{(\tau)}$ is a sum of copies of $C_1^{(\tau)}$, $(m_2 \otimes n_2)^{(\sigma)}$ a sum of copies of $C_2^{(\sigma)}$, $(m_2 \otimes n_2)^{(\tau)}$ a sum of copies of $C_2^{(\tau)}$. This and the preceding paragraph show that every irreducible A-module is a submodule of some $(m \otimes n)^{(\sigma)}$, $(m \otimes n)^{(\tau)}$. The last paragraph preceding these shows that the same conclusion applies to all irreducible $\mathcal{D}(\mathcal{B},*)$-modules.

With m and n as above, m becomes an irreducible right C-module $m^{(\sigma)}$ resp. $m^{(\tau)}$ by setting $m \cdot x = x^{\sigma}m$ resp. $x^{\tau}m$. Thus m has the structure of irreducible λ_n-admissible L_0-module, with

$$m(e_{n+1} \otimes b) = \frac{1}{2}\, m \cdot b = \left\{ \begin{array}{ll} \frac{1}{2}\, bm, & \text{in } m^{(\sigma)} \\ -\frac{1}{2}\, bm, & \text{in } m^{(\tau)} \end{array} \right\},$$

and with $mS_{b,c} = \frac{1}{4}[c,b]m$ in each case. Of course, n is such an L_0-module, with

$$n(e_{n+1} \otimes b) = \frac{1}{2}\, nb, \quad nS_{b,c} = \frac{1}{4}\, n[b,c].$$

It follows by §II.2 that $m^{(\sigma)} \otimes n$ and $m^{(\tau)} \otimes n$ are direct sums of irreducible $2\lambda_n$-admissible L_0-modules. The determination above show that the tensor-product structure of L_0-module on $m^{(\sigma)} \otimes n$ is the same as the structure of L_0-module on $m \otimes n$, regarded as the $\mathcal{D}(\mathcal{B},*)$-module $(m \otimes n)^{(\sigma)}$, obtained in §4; likewise with σ replaced by τ. We have seen that every $\mathcal{D}(\mathcal{B},*)$-module is a summand of one of the latter. That is, *every irreducible* $\mathcal{D}(\mathcal{B},*)$*-module is a* $2\lambda_n$*-admissible irreducible* L_0*-module; indeed, it is a submodule of the tensor product of two* λ_n*-admissible irreducible* L_0*-modules.*

(Essentially the only modification for r odd is the interchange of σ and τ throughout.)

In the language of §V.7, we now have the sharp description of the fundamental set given below. The notations are those of §1.

THEOREM VII.1. *A fundamental set for* L *consists of the weights* $\lambda_1, \ldots, \lambda_n$*, and for weights* μ *in this set of* μ*-admissible irreducible* L_0*-modules* $\mathcal{W}^{(\mu)}$ *as follows:*

For $\mu = \lambda_i$, $i < n$; $\mathcal{W}^{(\mu)} = F$, *as in §2; there is one module for each* λ_i.

For $\mu = \lambda_n$: *There is one module, afforded as in §3 by a minimal right ideal of the Clifford algebra* $C\ell(\mathcal{B},*)$*, except when* q *is odd and* $C\ell(\mathcal{B},*)$ *is not simple; in the exceptional case, there are two modules, afforded as in §3 by the two inequivalent irreducible* $C\ell(\mathcal{B},*)$*-modules.*

CHAPTER VIII

EXCEPTIONAL TYPES. I: F_4 WITH ASSOCIATIVE COEFFICIENTS.

In §III.5 of [26], it is shown that a central simple Lie algebra L of relative type F_4 over F results from what it known as "Tits' second construction", involving the split exceptional Jordan algebra and a composition division algebra. The latter must be one of the following:

i) F1; then L is split, and of no further interest here.

ii) a quadratic field extension K of F.

iii) a quaternionic division algebra $\mathcal{Q}$ over F.

iv) an octonionic division algebra $\mathcal{O}$ over F.

In this chapter we treat the cases ii) and iii), reserving iv) for the next chapter. Some of the techniques used in dealing with iii) involve an embedding in L of an algebra treated in Chapter V. These methods will be applied more extensively in the next chapter, to a pair of embeddings $\mathfrak{sl}(3,\mathcal{O}) \hookrightarrow \mathfrak{sp}(6,\mathcal{O}), \mathfrak{sl}(3,\mathcal{O}) \hookrightarrow F_4(\mathcal{O})$, where $\mathcal{O}$ is octonionic.

1. DECOMPOSITION OF THE LIE ALGEBRAS. FUNDAMENTAL WEIGHTS.

Let $\mathcal{D}$ be one of the composition algebras i)-iv). Let J be the split exceptional simple Jordan algebra over F, explicitly realized as the 3 by 3 hermitian matrices over the split octonion algebra C. We have in J the composition

$$x\cdot y = xy + yx,$$

juxtaposition denoting ordinary multiplication of 3 by 3 C-matrices. The diagonal of a matrix $x \in J$ has scalar entries, so its trace, Tr(x), is in F. Set $(x,y) = \mathrm{Tr}(x\cdot y)$ and $x\circ y = xy + yx - \frac{1}{3}(x,y)\, I \in J_0$, the kernel of Tr. All derivations of J map J into J_0, and are linear combinations of the "inner" derivations of the form

$$D_{x,y}\colon z \mapsto (z\cdot x)\cdot y - (z\cdot y)\cdot x.$$

Der(J) is the split simple Lie algebra of type F_4 over F.

The Lie algebra $L = F_4(\mathcal{D})$ is the vector space

$$L = \mathrm{Der}(\mathcal{D}) \oplus \mathrm{Der}(J) \oplus (J_0 \otimes \mathcal{D}_0),$$

where $\mathcal{D}_0$ is the kernel of the linear mapping $\tau\colon\mathcal{D} \to F$ determined by the fact that every $a \in \mathcal{D}$ satisfies a quadratic equation $a^2 - 2\tau(a)a + n(a) = 0$, where τ is linear and n quadratic with $\tau(1) = 1 = n(1)$. (If $\mathcal{D} = F$, $\tau(a) = a$ while $n(a) = a^2$; thus $\mathcal{D}_0 = 0$ in this case.) For $x,y \in \mathcal{J}_0$; $a, b \in \mathcal{D}_0$, we have

$$[x \otimes a, y \otimes b] = \frac{(x,y)}{6} D_{a,b} + \frac{ab+ba}{2} D_{x,y} + (x \circ y) \otimes \frac{[ab]}{2},$$

where the "quadratic equation " in $\mathcal{D}$ shows that $ab + ba \in F1$ if $a,b \in \mathcal{D}_0$, and where $[ab] = ab - ba \in \mathcal{D}_0$ for all $a, b \in \mathcal{D}$. The derivation $D_{a,b}$ of $\mathcal{D}$ is the "inner derivation":

$$c \mapsto (ca)b - (cb)a + b(ac) - a(bc) + (ac)b - a(cb)$$

$$= c[R_a, R_b] + c[L_a, L_b] + c[L_a, R_b],$$

where L_a is left multiplication by a, R_a is right. Of course $\mathrm{Der}(\mathcal{D}) = 0$ in cases i) and ii); in case iii), $D_{a,b} = \mathrm{ad}([ab])$. Again the inner derivations $D_{a,b}$ span $\mathrm{Der}(\mathcal{D})$.

For a decomposition of L into root-spaces, we refer to [13] and [24] very freely, making only the following modifications:

a) the basis $u_1, \ldots, u_8$ for C used in [24] (p. 287) is to be relabeled by making the switches $u_5 \leftrightarrow u_8$, $u_6 \leftrightarrow u_7$;
b) the formulas (6), (7), (8) of [24] are incorrect: each scalar product (a, a_{12}), (a, a_{13}), (a, a_{23}) should be multiplied by 2.

A maximal split torus T in L, $T \subseteq \mathrm{Der}(\mathcal{J})$, has as basis

$H_1 = 2(0,0,0, E_{33} - E_{22} - E_{66} + E_{77})$,

$H_2 = 2(0,0,0, - E_{11} - E_{33} + E_{66} + E_{88})$,

$H_3 = 4(0,0,0, E_{11} - E_{88})$

$H_4 = 2(0,0,0, - E_{11} + E_{22} + E_{33} + E_{44} - E_{55} - E_{66} - E_{77} + E_{88})$.

A fundamental system of roots $\alpha_1, \ldots, \alpha_4$ relative to T has as corresponding root-spaces:

$$L_{\alpha_1} = F\cdot(0,0,0, E_{23} - E_{67}),$$

$$L_{\alpha_2} = F\cdot(0,0,0, E_{16} - E_{38}),$$

$$L_{\alpha_3} = F\cdot(u_1,0,0,0) + u_1(1,2) \otimes \mathcal{D}_0,$$

$$L_{\alpha_4} = F\cdot(0,u_8,0,0) + u_8(1,3) \otimes \mathcal{D}_0.$$

Thus

$$L_{-\alpha_1} = F\cdot(0,0,0,\ E_{32} - E_{76}),$$

$$L_{-\alpha_2} = F\cdot(0,0,0,\ E_{61} - E_{83}),$$

$$L_{-\alpha_3} = F\cdot(u_8,0,0,0) + u_8(1,2) \otimes \mathcal{D}_0,$$

$$L_{-\alpha_4} = F\cdot(0,u_1,0,0) + u_1(1,3) \otimes \mathcal{D}_0.$$

The diagram for $\alpha_1, \alpha_2, \alpha_3, \alpha_4$ is of type F_4:

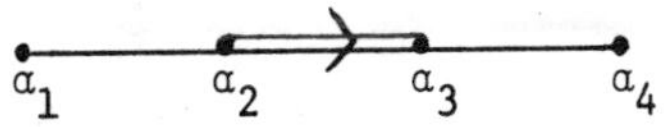

and $\alpha_i(H_j) = A_{ij}$, $1 \le i, j \le 4$, as before.

When $\lambda_1, \lambda_2, \lambda_3, \lambda_4$ is the basis for T^* dual to H_1, H_2, H_3, H_4, we see from splitting L that the restrictions to T of the fundamental weights relative to a split Cartan subalgebra H containing T, and a compatible ordering of the roots relative to H (Appendix 5) are:

$$\{\lambda_1, \lambda_2, \lambda_3, \lambda_4\} \text{ if } \mathcal{D} = F \text{ or } \mathcal{D} = K\ ;$$

$$\{\lambda_1, \lambda_2, \lambda_3, \lambda_4, 2\lambda_3, \lambda_3 + \lambda_4, 2\lambda_4\}, \text{ if } \mathcal{D} = \mathcal{Q}\ ;$$

$$\{\lambda_1, \lambda_2, 2\lambda_3, 2\lambda_4, \lambda_3 + \lambda_4, 2\lambda_3 + \lambda_4, \lambda_3 + 2\lambda_4, 2\lambda_3 + 2\lambda_4\},$$

$$\text{if } \mathcal{D} = \mathcal{O}\ .$$

Furthermore, when $\mathcal{D} = \mathcal{Q}$, there are only two irreducible representations in the split case whose highest weights have restriction $2\lambda_3$ to T, and likewise for $\lambda_3 + \lambda_4$ and $2\lambda_4$. When $\mathcal{D} = \mathcal{O}$, there is only one such for each of the weights in the corresponding list above, except for $2\lambda_3 + 2\lambda_4$, where there are three.

The centralizer L_0 of T is described in the terms of [26] as

$$L_0 = \mathrm{Der}(\mathcal{D}) \oplus T \oplus (U_3 \otimes \mathcal{D}_0) \oplus (U_4 \otimes \mathcal{D}_0),$$

where $U_3, U_4 \in J_0$ are the respective diagonal matrices $\mathrm{diag}\{-1,1,0\}$, $\mathrm{diag}\{-1,0,1\}$. In the notations as adapted, with corrections, we have from [24] that

$$[(u_8,0,0,0),\ (u_1,0,0,0)] = -\tfrac{1}{4} H_3;$$

$$[(0,u_1,0,0),\ (0,u_8,0,0)] = -H_4;$$

$$[u_8(1,2) \otimes \mathcal{D}_0,\ (u_1,0,0,0)] = U_3 \otimes \mathcal{D}_0\ ;$$

$$[u_1(1,3) \otimes \mathcal{D}_0,\ (0,u_8,0,0)] = U_4 \otimes \mathcal{D}_0\ ;$$

$$[u_8(1,2) \otimes a,\ u_1(1,2) \otimes b] =$$

$$\frac{1}{3} D_{a,b} + \frac{ab+ba}{2} H_3 + \frac{1}{3} U_3 \otimes \frac{[ab]}{2} - \frac{2}{3} U_4 \otimes \frac{[ab]}{2},$$

for $a, b \in \mathcal{D}_0$.

2. CONSTRUCTION OF IRREDUCIBLE REPRESENTATIONS: λ_1 AND λ_2.

If λ is one of λ_1, λ_2, and if W is an irreducible λ-admissible L_0-module, then W is annihilated by $[L_{-\alpha_3} L_{\alpha_3}]$ and by $[L_{-\alpha_4} L_{\alpha_4}]$. From our realizations of $L_{\pm\alpha_3}, L_{\pm\alpha_4}$, and from the last five formulas of §1 we see that W is annihilated by H_3, H_4, $U_3 \otimes \mathcal{D}_0$, $U_4 \otimes \mathcal{D}_0$ and all $D_{a,b}$ for $a, b \in \mathcal{D}_0$. As we have noted in §1, the $D_{a,b}$ for $a, b \in \mathcal{D}$, span $\mathrm{Der}(\mathcal{D})$; but each $a \in \mathcal{D}$ has the form $a = \tau(a)1 + a$, where $\tau(a) \in F$, $a_0 \in \mathcal{D}_0$, and $D_{1,b} = D_{b,1} = 0$ for all b. Thus $D_{a,b} = D_{a_0,b_0}$, and W is annihilated by $\mathrm{Der}(\mathcal{D})$. Finally, if $\lambda = \lambda_1$, W is annihilated by H_2, and likewise with the indices 1 and 2 interchanged. Thus the action on $w \in W$ of

$$D + \sum_{i=1}^{4} \mu_i H_i + (U_3 \otimes a_0) + (U_4 \otimes b_0),$$

where $D \in \mathrm{Der}(\mathcal{D})$, $\mu_i \in F$; $a_0, b_0 \in \mathcal{D}_0$, is scalar multiplication by μ_1 resp. μ_2 according as $\lambda = \lambda_1$ or λ_2.

It follows that there is at most one irreducible λ-admissible L_0-module for each of these choices of λ; the module is one-dimensional, with the action prescribed above. One will easily verify that a one-dimensional space with that respective action of L_0 is an irreducible λ-admissible L_0-module. Thus there is a unique irreducible λ-admissible L_0-module for $\lambda = \lambda_1$ or λ_2; it is one-dimensional with action as above and is therefore absolutely irreducible. It follows that the irreducible L-module with this L_0-module as subspace annihilated by N is absolutely irreducible.

The module of highest weight λ_1 is the adjoint module, of dimension 78, 133 or 248 according as $\mathcal{D} = K$, $\mathcal{Q}$ or $\mathcal{O}$. From the splitting data of Appendix 5, we see that upon splitting the module of highest weight λ_2 becomes that of highest weight π_3, π_5 or π_2 as $\mathcal{D} = K$, $\mathcal{Q}$ or $\mathcal{O}$. The respective dimensions are: 2925; 8645; 30,380.

3. THE CASE $\mathcal{D} = K$: λ_3 AND λ_4.

The situation for λ_3 is typical. If W is a λ_3-admissible irreducible L_0-module, W must be annihilated by H_1, H_2, $[L_{-\alpha_4}, L_{\alpha_4}]$, and by the projection on $U(L_0)$ of $e^2_{-\alpha_3} e^2_{\alpha_3}$ for all $e_{\pm\alpha_3}$. It follows that H_4

and $U_4 \otimes K_0$ annihilate $\mathcal{W}$, and that

$$w\psi((u_8,0,0,0)^2\ (u_1(1,2) \otimes a_0)^2) = 0$$

for all $w \in \mathcal{W}$, $a_0 \in K_0$. Expanding this last and using the commutation rules in L gives

$$w(U_3 \otimes a_0)^2 = w(D_{u_8(1,2),\ u_1(1,2)} \otimes a_0^2)$$

But $D_{u_8(1,2),u_1(1,2)} = H_3$, and $a_0^2 \in F$; thus

$$w(U_3 \otimes a_0)^2 = a_0^2 w \quad \text{for all} \quad w \in \mathcal{W},\ a_0 \in K_0.$$

It follows that if we set $wa = \tau(a)w + w(U_3 \otimes a_0)$, then $\mathcal{W}$ becomes a right K-module, necessarily irreducible, hence a one-dimensional K-vector space, its structure of L_0-module being given by

(1) $$w(\Sigma\mu_i H_i + U_3 \otimes a_0 + U_4 \otimes b_0) = w(\mu_3 + a_0).$$

Likewise for λ_4, we obtain $\mathcal{W} = K$, the action (1) being replaced by $w(\mu_4 + b_0)$.

With $\mathcal{D} = K$, these are readily seen to be λ-admissible irreducible L_0-modules for $\lambda = \lambda_3$ or λ_4. Upon an extension E splitting K (say by K itself), each splits into two inequivalent one-dimensional $(L_0)_E$-modules. These latter are then absolutely irreducible and correspond, in our labeling of Appendix 5, to the split highest weights

$$\pi_2 \text{ and } \pi_4, \text{ for } \lambda_3;$$

$$\pi_1 \text{ and } \pi_5, \text{ for } \lambda_4.$$

The corresponding irreducible L-modules are <u>not</u> absolutely irreducible, and have dimensions:

For λ_4: 54; for λ_3 : 702.

4. AN EMBEDDING OF $s\ell(3,\mathcal{Q})$ IN $F_4(\mathcal{Q})$.

Let M be the subalgebra of L generated by $L_{\pm\alpha_3}$ and $L_{\pm\alpha_4}$, where we assume $\mathcal{D} = \mathcal{Q}$. Much of this section applies as well when $\mathcal{D} = \mathcal{O}$, a case that will be invoked in Chapter IX. First we note that M <u>is simple and that</u> $\mathcal{U} = FH_3 + FH_4$ <u>is a maximal split torus in</u> M.

For the definitions of $L_{\pm\alpha_3}$ and $L_{\pm\alpha_4}$ in §1 and the formulas at the end of that section show that M contains H_3, H_4, $U_3 \otimes \mathcal{D}_0$, $U_4 \otimes \mathcal{D}_0$ and all $D_{a,b}$, hence all of $\mathrm{Der}(\mathcal{D})$, as in §2. Thus $M \cap L_0$ is of codimension at most 2 in L_0. Evidently $M_0 = \mathrm{Der}(\mathcal{D}) + FH_3 + FH_4 + U_3 \otimes \mathcal{D}_0 + U_4 \otimes \mathcal{D}_0$ is a subalgebra of $M \cap L_0$, containing $[L_{\alpha_3} L_{-\alpha_3}]$, $[L_{\alpha_4} L_{-\alpha_4}]$ and (from $[L_{\alpha_3} L_{\alpha_4}] = L_{\alpha_3+\alpha_4}$ and $[L_{-\alpha_3} L_{-\alpha_4}] = L_{-(\alpha_3+\alpha_4)})[L_{\alpha_3+\alpha_4}, L_{-(\alpha_3+\alpha_4)}]$. It follows that M is the sum of $M_0, L_{\pm\alpha_3}, L_{\pm\alpha_4}, L_{\pm(\alpha_3+\alpha_4)}$ and that M_0 is the centralizer of U in M. Thus $L_0 = FH_1 \oplus FH_2 \oplus M_0$, and therefore M_0 acts irreducibly on each root-space.

Moreover, $[L_0 L_0] = \mathrm{Der}(\mathcal{D}) + U_3 \otimes \mathcal{D}_0 + U_4 \otimes \mathcal{D}_0 = [M_0 M_0]$ contains no elements $x \neq 0$ such that ad x is diagonalizable in F (either in the action on L or in that on $[M_0, M_0]$, the two being equivalent by the semisimplicity of $[M_0 M_0]$). Thus U is a maximal split torus in M, since its centralizer in M is $M_0 = U + [M_0 M_0]$.

To see that M is simple, note that since U stabilizes any ideal I in M, I must be the sum of $I \cap M_0$ and of the intersections of I with the six root-spaces. As soon as $I \cap L_\alpha \neq 0$ for some α, the irreducibility of L_α as M_0-module yields $L_\alpha \subseteq I$. In this case it is an easy matter to see that $I = M$. If $I \subseteq M_0$, then $[I L_\alpha] = 0$ for all six roots α, from which I is <u>central</u> in M. Forming brackets with elements of $\mathrm{Der}(\mathcal{D}) \subseteq M_0$ and using the simplicity of $\mathrm{Der}(\mathcal{D})$ and its irreducibility on $\mathcal{D}_0$, we have $I \subseteq U$. But no non-zero element of U is annihilated by both α_3 and α_4; thus $I = 0$, and M is simple.

Rather than writing out an explicit isomorphism from $\mathfrak{sl}(3,\mathcal{D})$ onto M, we define one on the centralizer $\mathfrak{sl}(3,\mathcal{D})_0$ of a maximal split torus in $\mathfrak{sl}(3,\mathcal{D})$ and use the theorem of Allison ([2], Theorem 9.1; [26], p. 132) to show that this map admits a suitable extension to the desired isomorphism. For the rest of this chapter, we consider only the case $\mathcal{D} = \mathcal{Q}$; that of $\mathcal{D} = \mathcal{O}$, with an appropriate definition of $\mathfrak{sl}(3,\mathcal{O})$, will be taken up in the next chapter.

Now $\mathfrak{sl}(3,\mathcal{Q})_0$ is the set of diagonal matrices in $\mathfrak{sl}(3,\mathcal{Q})$, each of which can be written $cI + H_1 \otimes a + H_2 \otimes b$, where $H_1 = \mathrm{diag}\ \{-1,1,0\}$; $H_2 = \mathrm{diag}\ \{0,-1,1\}$; I is the identity matrix; $a,b \in \mathcal{Q}$; $c \in [\mathcal{Q}\mathcal{Q}]$. We may identify c with the inner derivation $x \mapsto [xc]$ of $\mathcal{Q}$, obtaining an isomorphism of Lie algebras between $[\mathcal{Q}\mathcal{Q}]$ and $\mathrm{Der}\ \mathcal{Q}$. Define a map $\theta_0 \colon \mathfrak{sl}(3,\mathcal{Q})_0 \to M_0$ by

$$D + H_1 \otimes a + H_2 \otimes b \mapsto$$
$$D + \tau(a)H_4 + \tau(b)H_3 + U_3 \otimes b_0 - U_4 \otimes a_0,$$

where $D \in \mathrm{Der}(\mathcal{Q})$, as in the identification above. This is evidently an isomorphism of linear spaces, and one checks routinely that it is an isomorphism of Lie algebras.

To establish the desired isomorphism of $\mathfrak{sl}(3,\mathcal{Q})$ onto M, we must next verify a condition of compatibility with θ_0 for representatives of the elements of the two Weyl groups (see [26], Chapter IV). Both Weyl groups are isomorphic to the symmetric group S_3. In $\mathfrak{sl}(3,\mathcal{Q})$, we represent the reflection w_{α_1} by the automorphism $\omega(w_{\alpha_1}) = \exp(\mathrm{ad}E_{12})\exp(-\mathrm{ad}E_{21})\exp(\mathrm{ad}E_{12})$, and take $\omega(w_{\alpha_2}) = \exp(\mathrm{ad}E_{23})\exp(-\mathrm{ad}E_{32})\exp(\mathrm{ad}E_{23})$. Each of these stabilizes $\mathfrak{sl}(3,\mathcal{Q})_0$ and fixes $\mathrm{Der}(\mathcal{Q})$; $\omega(w_{\alpha_1})$ sends $H_1 \otimes a$ to $-H_1 \otimes a$ and $H_2 \otimes b$ to $H_1 \otimes b + H_2 \otimes b$; $\omega(w_{\alpha_2})$ sends $H_1 \otimes a$ to $H_1 \otimes a + H_2 \otimes a$, $H_2 \otimes b$ to $-H_2 \otimes b$. Then elements of the Weyl group are: id, w_{α_1}, w_{α_2}, $w_{\alpha_1}w_{\alpha_2}$, $w_{\alpha_2}w_{\alpha_1}$, $w_{\alpha_1}w_{\alpha_2}w_{\alpha_1}$. We assign the identity automorphism as representative of "id", and the products of the corresponding $\omega(w_{\alpha_i})$ to the other elements. Thus, for example, $w_{\alpha_1+\alpha_2} = w_{\alpha_1}w_{\alpha_2}w_{\alpha_1}$ has representative $\omega(w_{\alpha_1})\omega(w_{\alpha_2})\omega(w_{\alpha_1})$.

Likewise, we represent the reflection w_{α_4} in the Weyl group of M by

$\omega(w_{\alpha_4}) = \exp(\mathrm{ad}(0,u_8,0,0))\exp(\mathrm{ad}\ 2(0,u_1,0,0))\exp(\mathrm{ad}(0,u_8,0,0))$,

which fixes $D \in \mathrm{Der}(\mathcal{Q})$, sends H_4 to $-H_4$, H_3 to $H_3 \pm H_4$, $U_3 \otimes a_0$ to $U_3 \otimes a_0 - U_4 \otimes a_0$, $U_4 \otimes a_0$ to $-U_4 \otimes a_0$; and the reflection w_{α_3} by

$\omega(w_{\alpha_3}) = \exp(\mathrm{ad}(u_1,0,0,0))\exp(\mathrm{ad}\ 2(u_8,0,0,0))\exp(\mathrm{ad}(u_1,0,0,0))$,

fixing $D \in \mathrm{Der}(\mathcal{Q})$, sending H_3 to $-H_3$, H_4 to $H_3 + H_4$, $U_3 \otimes a_0$ to $-U_3 \otimes a_0$, $U_4 \otimes a_0$ to $U_4 \otimes a_0 - U_3 \otimes a_0$. We then proceed as above to assign representatives, assigning $\omega(w_{\alpha_4})\omega(w_{\alpha_3})\omega(w_{\alpha_4})$ as representative for $w_{\alpha_4}w_{\alpha_3}w_{\alpha_4} = w_{\alpha_3}w_{\alpha_4}w_{\alpha_3} = w_{\alpha_3+\alpha_4}$.

The isomorphism θ_0 maps the maximal split torus $FH_1 + FH_2 \subseteq \mathfrak{sl}(3,\mathcal{Q})$ to $\mathcal{U}$, and the induced map on the dual spaces is easily seen to send α_4 to α_1, α_3 to α_2. There results an isomorphism of the Weyl groups, in which w_{α_1} and w_{α_4} correspond, as do w_{α_3} and w_{α_2}. In the formulation of Allison's theorem given in [26], we must consider the automorphisms of $\mathfrak{sl}(3,\mathcal{Q})_0$ of the form

(2) $$\theta_0\, \omega'\, \theta_0^{-1}\, \omega^{-1},$$

where ω runs over our chosen set of representatives for the elements of the Weyl group of $\mathfrak{sl}(3,\mathcal{Q})$ and where ω' is the chosen representative for the corresponding element in the Weyl group of M. But now it is an easy matter to verify that (2) is the identity mapping of $\mathfrak{sl}(3,\mathcal{Q})_0$ when $\omega = \omega(w_{\alpha_1})$ and

$\omega' = \omega(w_{\alpha_4})$, as well as when $\omega = \omega(w_{\alpha_2})$ and $\omega' = \omega(w_{\alpha_3})$. That (2) is the identity for all elements of the Weyl group follows by our choice of representatives as corresponding products.

For the mappings (2) to be "admissible", so that Allison's theorem applies, they are required to fix certain subspaces of $\mathfrak{sl}(3,\mathcal{Q})_0$ and to have the property that the representations of $\mathfrak{sl}(3,\mathcal{Q})_0$ on certain subspaces of $\mathfrak{sl}(3,\mathcal{Q})$ shall be equivalent with the same representation "twisted" by the automorphism (2) of $\mathfrak{sl}(3,\mathcal{Q})_0$. When all the automorphisms (2) are the identity, these last conditions are trivially satisfied.

The final verification for the application of Allison's theorem is to display an isomorphism of $\mathfrak{sl}(3,\mathcal{Q})_0$-modules from one root-space in $\mathfrak{sl}(3,\mathcal{Q})$ to a corresponding one in M, using the map θ_0 to make the latter root-space into an $\mathfrak{sl}(3,\mathcal{Q})_0$-module. We treat the space $E_{23} \otimes \mathcal{Q}$ in $\mathfrak{sl}(3,\mathcal{Q})$, corresponding to α_2, and map it onto $L_{\alpha_3} \subseteq M$ by mapping

$$E_{23} \otimes a \quad \text{to} \quad 2\tau(a)(u_1,0,0,0) + u_1(1,2) \otimes a_0.$$

What must be checked is the equality of the image of

$$[E_{23} \otimes a,\ D + H_1 \otimes b + H_2 \otimes c]$$

(3) $$= E_{23} \otimes aD - E_{23} \otimes ba + E_{23} \otimes (ac + ca)$$

with

(4) $$[2\tau(a)(u_1,0,0,0) + u_1(1,2) \otimes a_0,$$
$$D + \tau(b)H_4 + \tau(c)H_3 + U_3 \otimes c_0 - U_4 \otimes b_0].$$

Working in $\mathrm{Der}(J)$ as in [24], one notes that

$$D_{u_1(1,2),U_4} = 2(u_1,0,0,0) \quad \text{and} \quad D_{u_1(1,2),U_3} = 4(u_1,0,0,0);$$

then the rules for Tits' second construction yield that (4) is equal to

$$(4\tau(a)\tau(c) - 2\tau(a)\tau(b) + 2(a_0c_0 + c_0a_0) - (a_0b_0 + b_0a_0))(u_1,0,0,0)$$
$$+ u_1(1,2) \otimes (aD - \tau(b)a_0 - \tau(a)b_0 + \frac{[a_0b_0]}{2} + 2\tau(a)c_0 + 2\tau(c)a_0),$$

which is seen directly to be the image of (3) under our map from $E_{23} \otimes \mathcal{Q}$.

All the conditions of Allison's theorem, in the version of [26] (§IV.3), are now satisfied. It follows that θ_0 and the displayed map of $E_{23} \otimes \mathcal{Q}$ to L_{α_3} admit an extension to an isomorphism θ of $\mathfrak{sl}(3,\mathcal{Q})$ onto M. The map θ necessarily sends $E_{12} \otimes \mathcal{Q}$ to L_{α_4} and the respective transposes to $L_{-\alpha_3}$, $L_{-\alpha_4}$.

Now let $\lambda \in T^*$ be dominant integral and let W be an irreducible L_0-module of weight λ. By the homomorphism θ_0 of $s\ell(3,\mathcal{Q})_0$ into L_0, W may be regarded as a module for $s\ell(3,\mathcal{Q})_0$. Since $L_0 = \theta_0(s\ell(3,\mathcal{Q})_0) + FH_1 + FH_2$, with H_1 and H_2 acting as scalars and centralizing L_0, W is an irreducible $s\ell(3,\mathcal{Q})_0$-module of weight $\lambda \circ \theta_0$ on the canonical maximal split torus in $s\ell(3,\mathcal{Q})$. Conversely, if W is simultaneously a T-module of weight λ and an irreducible $s\ell(3,\mathcal{Q})_0$-module where $\lambda(\theta_0(T))$ agrees with the action of T for each T in the canonical maximal split torus of $s\ell(3,\mathcal{Q})$, then W is an irreducible L_0-module of weight λ. To say that W is λ-admissible under these circumstances is to say that for each i, $1 \le i \le 4$, and for each $e_{\alpha_i} \in L_{\alpha_i}$ and each $e_{-\alpha_i} \in L_{-\alpha_i}$, the projection of

$e_{-\alpha_i}^{\lambda(H_i)+1} e_{\alpha_i}^{\lambda(H_i)+1} \in U(L)$ on $U(L_0)$ in the decomposition

$$U(L) = U(L_0) + NU(N + L_0) \oplus U(L_0 + N^-)N^- \oplus NU(L)N^- \tag{5}$$

annihilates W.

By the fact that $L_{\pm\alpha_1}$, $L_{\pm\alpha_2}$ are one-dimensional, this last condition is automatic for $i = 1,2$ from representations of split 3-dimensional algebras. The injection θ of $s\ell(3,\mathcal{Q})$ into L is consistent with (5), in the sense that the corresponding projection of $U(s\ell(3,\mathcal{Q}))$ onto $U(s\ell(3,\mathcal{Q})_0)$ is the restriction of the projection of (5). Thus if W is λ-admissible as L_0-module, W is $(\lambda \circ \theta_0)$-admissible as $s\ell(3,\mathcal{Q})_0$-module and conversely.

5. REPRESENTATIONS OF $F_4(\mathcal{Q})$: HIGHEST WEIGHTS λ_3, λ_4 AND COMBINATIONS.

Applying the considerations of the last section, we see that the irreducible λ_3-admissible L_0-modules are the irreducible $\lambda_3 \circ \theta_0$-admissible $s\ell(3,\mathcal{Q})_0$-modules, and that $\lambda_3 \circ \theta_0$ is what was called " λ_2" in the treatment of general $s\ell(3,\mathcal{Q})$-modules in §V. In that treatment, we found a unique such module, W, with underlying space $\mathcal{Q}$ (for our setting), the action of $\operatorname{diag}\{\delta_1, \delta_2, \delta_3\}$ $(\delta_i \in \mathcal{Q}, \tau(\delta_1 + \delta_2 + \delta_3) = 0)$ on $w \in \mathcal{Q}$ being $w\delta_3$. The corresponding action of the general element $D + \sum_{i=1}^{4} \mu_i H_i + U_3 \otimes a_0 + U_4 \otimes b_0$ of L_0 on $w \in \mathcal{Q}$ is then seen, using θ_0, to be

$$\mu_3 w + w(c_0 + a_0),$$

where $D \in \operatorname{Der}(\mathcal{Q})$ is the mapping $x \to [xc_0]$ $(c_0 \in \mathcal{Q}_0)$. Evidently this W is irreducible, but splits into two isomorphic L_0-modules upon splitting $\mathcal{Q}$.

Then the latter are absolutely irreducible; the corresponding representation of a split L has highest weight π_7 and dimension 912. Our irreducible L-module of highest weight λ_3 thus has dimension 1824.

For $\lambda = \lambda_4$, the situation is as for λ_3, with $w\delta_3$ replaced by $-\delta_1 w$, so that

$$w(D + \Sigma\mu_i H_i + U_3 \otimes a_0 + U_4 \otimes b_0)$$

$$= \mu_4 w - (b_0 + c_0)w, \quad \text{in the previous notation.}$$

Again W (and the corresponding L-module) is unique but not absolutely irreducible. The L-module splits into two isomorphic modules of highest weight π_1 (over a splitting field), each of dimension 56. Thus our unique λ_4-admissible irreducible L-module has dimension 112.

We have seen in §V.7 that the modules for $s\ell(3,Q)$ corresponding to irreducible L-modules of highest weights $2\lambda_3$, $2\lambda_4$, $\lambda_3 + \lambda_4$ (weights $2\lambda_2$, $2\lambda_1$, $\lambda_1 + \lambda_2$, in the setting of $s\ell(3,Q)$) arise by Cartan multiplication from those of highest weights: λ_2 and λ_2; λ_1 and λ_1; λ_1 and λ_2, respectively. From §3 above it now follows that there are two irreducible L-modules of each of the highest weights $2\lambda_3$, $2\lambda_4$, and one of highest weight $\lambda_3 + \lambda_4$. All of these result from the modules of highest weights λ_3 and λ_4 by Cartan multiplication.

6. SUMMARY.

THEOREM VIII.1. *Let K be a quadratic extension field of F, and let L be the central simple Lie algebra over F of relative type F_4 resulting from K and the split exceptional Jordan algebra by Tits' second construction. In the notations of §1 and with the conventions of §V.7, a fundamental set for L consists of the weights λ_1, λ_2, λ_3, λ_4, and, for each of these weights of a single irreducible admissible L_0-module W.*

For λ_1, λ_2, $W = F$, with the action of $g = \Sigma\mu_i H_i + (U_3 \otimes a_0) + (U_4 \otimes b_0)$ being multiplication by μ_1 resp. μ_2. For λ_3, λ_4, $W = K$, and g acts as multiplication by $\mu_3 + a_0$ resp. $\mu_4 + b_0$.

THEOREM VIII.2. *Let Q be a quaternionic division algebra over F, and let L result from Q and the split exceptional Jordan algebra by Tits' second construction. Then a fundamental set for L consists of the weights λ_1, λ_2, λ_3, λ_4, and, for each of these, of a single irreducible admisiible L_0-module W.*

For λ_1, λ_2, $W = F$, the action of

$$g = D + \Sigma\mu_i H_i + U_3 \otimes a_0 + U_4 \otimes b_0 \in L_0$$

being multiplication by μ_1 resp μ_2.

For λ_3, $\mathcal{W} = \mathcal{Q}$, *the action of* g *on* $w \in \mathcal{Q}$ *sending* w *to* $\mu_3 w + w(c_0 + a_0)$, *where* $D \in \mathrm{Der}(\mathcal{Q})$ *sends* x *to* $[x, c_0]$ $(c_0 \in \mathcal{Q}_0)$.

For λ_4, $\mathcal{W} = \mathcal{Q}$, *the action of* g *on* $w \in \mathcal{Q}$ *sending* w *to* $\mu_4 w - (b_0 + c_0)w$.

CHAPTER IX

EXCEPTIONAL TYPES II: LIE ALGEBRAS COORDINATIZED BY OCTONIONS.

The Lie algebras $\mathfrak{sl}(3,\mathcal{O})$, $\mathfrak{sp}(6,\mathcal{O})$, $F_4(\mathcal{O})$ whose representations are considered here are those resulting from Tits' second construction applied to an octonionic division algebra $\mathcal{O}$ and a split simple cubic Jordan algebra that is, respectively: $M_3(F)$; the fixed elements of $M_6(F)$ under an involution determined by a symplectic form; the split exceptional algebra J of Chapter VIII. They are of relative types A_2, C_3, F_4 and combine with the algebras of Chapters V, VI, VIII to exhaust all central simple algebras over F of these types [26]. Upon extension of F to its algebraic closure (or to an extension splitting $\mathcal{O}$-thus a quadratic extension will do), they become the split exceptional Lie algebras of types E_6, E_7, E_8.

1. THE ALGEBRA $\mathfrak{sl}(3,\mathcal{O})$; FUNDAMENTAL WEIGHTS.

The Lie algebra $\mathfrak{sl}(3,\mathcal{O})$ is constructed as the vector space $L = \mathrm{Der}(\mathcal{O}) \oplus (\mathfrak{sl}(3,F) \otimes \mathcal{O})$, with multiplication such that $\mathrm{Der}(\mathcal{O})$ is a subalgebra,

$$[x \otimes a, D] = x \otimes aD \quad (x \in \mathfrak{sl}(3,F),\ a \in \mathcal{O},\ D \in \mathrm{Der}(\mathcal{O})),$$

and

$$[x \otimes a,\ y \otimes b] = \frac{1}{3}\,\mathrm{Tr}(xy)D_{a,b} + [xy] \otimes \frac{ab+ba}{2} + (xy + yx - \frac{2}{3}\,\mathrm{Tr}(xy)I) \otimes \frac{[ab]}{2},$$

where $D_{a,b}$ is the inner derivation of $\mathcal{O}$ determined by a and b, as in §VIII.1. If $H_1 = E_{22} - E_{11} \in \mathfrak{sl}(3,F) \subseteq \mathfrak{sl}(3,F) \otimes \mathcal{O}$ and $H_2 = E_{33} - E_{22}$, then H_1 and H_2 span a maximal split torus T in L, whose centralizer is

$$L_0 = \mathrm{Der}\ (\mathcal{O}) + H_1 \otimes \mathcal{O} + H_2 \otimes \mathcal{O}\ .$$

The root-spaces relative to T are the spaces $E_{ij} \otimes \mathcal{O}$, $i \neq j$. We set $L_{\alpha_1} = E_{12} \otimes \mathcal{O}$, $L_{\alpha_2} = E_{23} \otimes \mathcal{O}$; then $L_{\alpha_1+\alpha_2} = E_{13} \otimes \mathcal{O}$, and the root-spaces for the negatives of these roots are obtained by taking transposes of the matrix units. The roots α_1 and α_2 form a fundamental system of type A_2; as before we let λ_1, λ_2 be the dual basis to H_1, H_2 in T^*, so that $\alpha_1 = 2\lambda_1 - \lambda_2$, $\alpha_2 = -\lambda_1 + 2\lambda_2$.

In Appendix 6, we shall see how a Cartan subalgebra H of L containing T splits under field extension, so that two of a fundamental set of (six) roots restrict to α_1 and α_2, the rest to zero on T. The corresponding restrictions of the six fundamental dominant integral functions in the split case will be seen there to be

$$\{2\lambda_1, 2\lambda_2, \lambda_1 + \lambda_2, 2\lambda_1 + \lambda_2, \lambda_1 + 2\lambda_2, 2\lambda_1 + 2\lambda_2\} .$$

From inspection of these restrictions, it is clear that there are exactly three inequivalent irreducible representations of the split algebra whose highest weight has $2\lambda_1 + 2\lambda_2$ as restriction (viz, those of highest weights $\pi_1 + \pi_5$, $2\pi_6$, π_3), and there is exactly one for each of the other weights in the list above.

2. THE ALGEBRA sp(6,$\mathcal{O}$). FUNDAMENTAL WEIGHTS.

This algebra is constructed as

$$L = \mathrm{Der}(\mathcal{O}) \oplus \mathrm{sp}(6,F) \oplus (M' \otimes \mathcal{O}\ )$$

where sp(6,F) is the split simple Lie algebra of type C_3, realized as the symplectic-skew transformations of a 6-dimensional vector space. The space M' is the 14-dimensional space of symplectic-symmetric transformations of trace zero in the same 6-dimensional space. The composition in L is given in Theorem III.7 of [26] and will be recalled as needed below.

With the standard involution $a \mapsto \bar{a}$ in $\mathcal{O}$, relative to which $\mathcal{O}_0$ is the space of skew elements, and with our 6-dimensional symplectic space having a symplectic basis $v_1, v_2, \ldots, v_6$ with $(v_1,v_6) = 1 = (v_2,v_5) = (v_3,v_4)$ and $(v_i,v_j) = 0$ if $i + j \neq 7$, the subspace

$$\mathrm{sp}(6,F) \oplus (M' \otimes \mathcal{O}_0)$$

of L may be viewed as the space of six-by-six $\mathcal{O}$-matrices (a_{ij}) satisfying:

$$\bar{a}_{ij} + a_{7-j,7-i} = 0,\ 1 \le i,\ j \le 3;$$

(1) $$a_{i,7-j} = \bar{a}_{j,7-i},\ a_{7-i,j} = \bar{a}_{7-j,i},\ 1 \le i,\ j \le 3;$$

$$s = \bar{s}, \quad \text{where} \quad s = \sum_{i=j}^{3} a_{ii}.$$

A maximal split torus T in L is the canonical one in sp(6,F), namely the linear span of the $E_{ii} - E_{7-i,7-i}$, $1 \le i \le 3$. The centralizer of T is

$$L_0 = \mathrm{Der}(\mathcal{O}) + T + U_1 \otimes \mathcal{O}_0 + U_2 \otimes \mathcal{O}_0,$$

where $U_1 = E_{22} - E_{11} + E_{55} - E_{66}$, $U_2 = E_{33} - E_{22} + E_{44} - E_{55}$, a basis for the subspace of M' centralizing T. A fundamental system of roots $\alpha_1, \alpha_2, \alpha_3$ relative to T has

$$L_{\alpha_1} = \{E_{12} \otimes a - E_{56} \otimes \bar{a} \mid a \in O\} ;$$

$$L_{\alpha_2} = \{E_{23} \otimes a - E_{45} \otimes \bar{a} \mid a \in O\} ;$$

$$L_{\alpha_3} = F \cdot E_{34} .$$

The corresponding $L_{-\alpha_i}$ are obtained by transposing the matrix-units. This system is of type C_3, with diagram

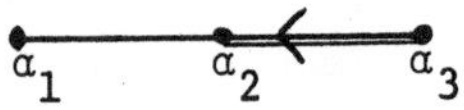

.

The related basis H_1, H_2, H_3 for T, with $\alpha_i(H_j)$ the Cartan integer A_{ij}, is:

$$H_1 = E_{22} - E_{11} - E_{55} + E_{66} ,$$

$$H_2 = E_{33} - E_{22} - E_{44} + E_{55} ,$$

$$H_3 = E_{44} - E_{33} .$$

We shall see in Appendix 6 that L splits to become the split exceptional simple algebra E_7 in such a way that the restrictions to T of the fundamental dominant integral functions $\pi_1, \dots, \pi_7$ for E_7 are

$$\{2\lambda_1,\ 2\lambda_2,\ \lambda_3,\ \lambda_1 + \lambda_2,\ 2\lambda_1 + \lambda_2,\ \lambda_1 + 2\lambda_2,\ 2\lambda_1 + 2\lambda_2\} ,$$

where, as usual, $(\lambda_1, \lambda_2, \lambda_3)$ is the basis for T^* dual to the basis (H_1, H_2, H_3) for T. Thus it is these weights for which we must construct all admissible L_0-modules.

[A corresponding description of $F_4(O)$, with a list of fundamental weights derived from Appendix 5, is given in §VIII.1.]

3. REPRESENTATIONS OF $s\ell(3,O)$: THE WEIGHTS $2\lambda_1$, $2\lambda_2$.

The two cases $\lambda = 2\lambda_1$, $\lambda = 2\lambda_2$ are essentially symmetric; we assume for now $\lambda = 2\lambda_2$, and let W be a λ-admissible irreducible L_0-module. Thus W is annihilated by

$$\psi(L_{-\alpha_1} L_{\alpha_1}) = [L_{-\alpha_1} L_{\alpha_1}]$$

so by all elements

$$[E_{21} \otimes a,\ E_{12} \otimes b] = \frac{1}{3} D_{a,b} + H_1 \otimes \frac{ab+ba}{2}$$
$$- \frac{2}{3} H_2 \otimes \frac{ab-ba}{2} - \frac{1}{3} H_1 \otimes \frac{ab-ba}{2}$$

for $a, b \in O$.

With $b = 1$, it follows from $D_{a,1} = 0$ that $H_1 \otimes O$ annihilates W. Thus for all $w \in W$; $a,b \in O$,

$$wD_{a,b} = w(H_2 \otimes (ab-ba)).$$

For $w \in \mathcal{W}$, $c \in \mathcal{O}$, let $wS_c = w(H_2 \otimes c)$. Then $S_1 = 2I$ and $wD_{a,b} = wS_{[ab]}$ by the above.

Now $[H_2 \otimes a, H_2 \otimes b] = \frac{2}{3} D_{a,b} + \frac{2}{3} H_1 \otimes [ab] + \frac{1}{3} H_2 \otimes [ab]$, so that $w[H_2 \otimes a, H_2 \otimes b] = wS_{[ab]}$. But since $\mathcal{W}$ is an L_0-module, $w[H_2 \otimes a, H_2 \otimes b] = w[S_a, S_b]$, and we have $[S_a, S_b] = S_{[ab]}$ for all $a,b \in \mathcal{O}$. The Jacobi identity for the endomorphisms S_a, S_b, S_c of $\mathcal{W}$ then yields

$$S_{[[ab]c]+[[bc]a]+[[ca]b]} = 0$$

for all $a,b,c \in \mathcal{O}$. It is readily verified (say, by checking the split case) that the linear combinations of elements $[[ab]c] + [[bc]a] + [[ca]b]$ constitute the subspace $\mathcal{O}_0$ of $\mathcal{O}$. Thus $S_c = 0$ for all $c \in \mathcal{O}_0$, and we have $wS_a = 2\tau(a)w$ for all $a \in \mathcal{O}$, $w \in \mathcal{W}$, where $a = \tau(a)1 + a_0$, $\tau(a) \in F$, $a_0 \in \mathcal{O}_0$. It follows that all $D_{a,b}$ (hence all elements of $\mathrm{Der}(\mathcal{O})$) annihilate $\mathcal{W}$. The effect on $w \in \mathcal{W}$ of the typical element of L_0,

$$g = D + H_1 \otimes a + H_2 \otimes b,$$

is to send w to $2\tau(b)w$.

Thus if there is a $2\lambda_2$-admissible irreducible L_0-module, it is unique and is a one-dimensional space with the action of $g \in L_0$ being given as above. Conversely, it is clear that such a space is an L_0-module of weight $2\lambda_2$, annihilated by all $\psi(e_{-\alpha_1} e_{\alpha_1})$. To see that it is in fact a $2\lambda_2$-admissible L_0-module, we must show only that it is annihilated by all $\psi(e^3_{-\alpha_2} e^3_{\alpha_2}) \in \mathcal{U}(L_0)$.

Now $[E_{32} \otimes a, E_{23} \otimes b] = \frac{1}{3} D_{a,b} + H_2 \otimes \frac{ab+ba}{2} + \frac{2}{3} H_1 \otimes \frac{[ab]}{2} + \frac{1}{3} H_2 \otimes \frac{[ab]}{2}$, so that in our module,

$$w\psi((E_{32} \otimes a)(E_{23} \otimes b)) = w[E_{32} \otimes a, E_{23} \otimes b]$$
$$= \tau(ab+ba)w = 2\,\tau(ab)w.$$

From the fundamental quadratic relation in $\mathcal{O}$,

$$a^2 - 2\tau(a)a + n(a)1 = 0, \text{ we have}$$
$$\tau(a^2) - 2\tau(a)^2 + n(a) = 0, \text{ so that for all } a \in \mathcal{O},$$

(2) $$a^2 - 2\tau(a)a + (2\tau(a)^2 - \tau(a^2))1 = 0.$$

Multiplying (2) by a, then applying τ, gives

(3) $$\tau(a^3) - 3\tau(a)\tau(a^2) + 2\tau(a)^3 = 0.$$

The required condition, the annihilation of $\mathcal{W}$ by all $\psi((E_{32} \otimes a)^3(E_{23} \otimes b)^3)$ in $\mathfrak{sl}(L_0)$, is a consequence of (3). For

$$\begin{aligned} &w\psi((E_{32} \otimes a)^3(E_{23} \otimes b)^3) \\ &= w\psi((E_{32} \otimes a)^2(E_{23} \otimes b)(E_{32} \otimes a)(E_{23}\otimes b)^2) \\ &+ w\psi((E_{32} \otimes a)^2[E_{32} \otimes a, E_{23} \otimes b]\ (E_{23} \otimes b)^2) \\ &= 3w\psi([E_{32}\otimes a, E_{23} \otimes b](E_{32} \otimes a)^2(E_{23} \otimes b)^2) \\ &+ 3w\psi((E_{32} \otimes a)[E_{32} \otimes a[E_{32} \otimes a, E_{23} \otimes b]](E_{23} \otimes b)^2). \end{aligned}$$

Now $[E_{32} \otimes a[E_{32} \otimes a, E_{23} \otimes b]]$ is seen from our rules of operation in $\mathfrak{sl}(3,\mathcal{O})$ to be

$$\frac{1}{2} E_{32} \otimes aD_{a,b} - 2E_{32} \otimes \tau(ab)a,$$

and $aD_{a,b} = a(ab+ba) + (ab+ba)a - 4aba$

$\qquad = 4\tau(ab)a - 4aba,$

(the subalgebra of $\mathcal{O}$ generated by a and b is associative) so that

$$\begin{aligned} &w\psi((E_{32} \otimes a)^3(E_{23} \otimes b)^3) \\ &= 6\psi(ab)w\psi((E_{32} \otimes a)^2(E_{23} \otimes b)^2) \\ &- 6w\psi((E_{32} \otimes a)(E_{32} \otimes aba)(E_{23} \otimes b)^2). \end{aligned}$$

Repetition of the step above gives

$$\begin{aligned} &w\psi((E_{32} \otimes a)^3(E_{23} \otimes b)^3) \\ &= (12\tau(ab)^3 - 18\tau(ab)\tau((ab)^2) + 6\tau((ab)^3))w, \end{aligned}$$

and this is zero by (3).

Thus a one-dimensional space $\mathcal{W}$, with the action of g being multiplication by $2\tau(b)$, is the unique $2\lambda_2$-admissible L_0-module. Likewise, a one-dimensional space on which g acts by scalar multiplication by $2\tau(a)$ is the unique $2\lambda_1$-admissible L_0-module.

The corresponding irreducible representations of L are those as norm-skew transformations of the exceptional simple Jordan algebra $H(M_3(\mathcal{O}))$ of [17], §1.5, and the contragredient of this representation. Both are absolutely irreducible of dimension 27 (e.g., see [13]).

4. REPRESENTATIONS OF $\mathfrak{sl}(3,\mathcal{O})$: THE WEIGHT $\lambda_1 + \lambda_2$.

This highest weight evidently occurs in the adjoint module. From remarks in §1, there is only one irreducible module of this highest weight (since it is absolutely irreducible), and we have it. Nevertheless, we present a direct approach, both to show the feasibility of "rational" methods and to obtain some relations for later use.

Let $\mathcal{W}$ be an irreducible $\lambda_1 + \lambda_2$-admissible L_0-module. We define $wS_a = w(H_2 \otimes a)$ as in §3, and $wT_a = w(H_1 \otimes a)$. Then $S_1 = I = T_1$. The fact that $\mathcal{W}$ is an L_0-module, applied using $H_2 \otimes a$ and $H_2 \otimes b$ as in §3, now yields

$$w[S_a,S_b] = \frac{2}{3}\, wD_{a,b} + \frac{2}{3}\, wT_{[ab]} + \frac{1}{3}\, wS_{[ab]} \;. \tag{4}$$

Similarly, using $H_1 \otimes a$ and $H_1 \otimes b$, then $H_1 \otimes a$, $H_2 \otimes b$, we have

$$w[T_a,T_b] = \frac{2}{3}\, wD_{a,b} - \frac{1}{3}\, wT_{[ab]} - \frac{2}{3}\, wS_{[ab]} \;, \tag{5}$$

$$w[T_a,S_b] = -\frac{1}{3}\, wD_{a,b} - \frac{1}{3}\, wT_{[ab]} + \frac{1}{3}\, wS_{[ab]} \;. \tag{6}$$

The condition of $\lambda_1 + \lambda_2$-admissibility adds the following constraints:

$$w\psi((E_{21} \otimes a)^2(E_{12} \otimes b)^2) = 0 = w\,((E_{32} \otimes a)^2(E_{23} \otimes b)^2).$$

With $b = 1$ these yield, respectively,

$$(T_a)^2 = T_{a^2},\; (S_a)^2 = S_{a^2} \;. \tag{7}$$

Polarizing the second of these gives

$$S_{ab+ba} = S_aS_b + S_bS_a \;. \tag{8}$$

Adding (4) to twice (6) gives

$$S_{[ab]} = [S_aS_b] + 2[T_a,S_b], \tag{9}$$

and addition of (8) and (9) yields

$$S_{ab} = S_aS_b + [T_a,S_b] \;. \tag{10}$$

Similarly, from (5), (6) and the first relation (7),

$$T_{ab} = T_bT_a + [S_b,\, T_a]. \tag{11}$$

The conditions (7), (10), (11) and the requirement that $T_1 = I = S_1$ are the conditions that $\mathcal{W}$ be an <u>alternative</u> <u>unital</u> $\mathcal{O}$ -<u>bimodule</u>, in the sense of [17], (p. 86). From any of (4)-(6) it is clear that, under all these conditions, any $\mathcal{O}$-subbimodule is an L_0-submodule, so that $\mathcal{W}$ must be an <u>irreducible</u> $\mathcal{O}$-bimodule.

From [16] (see also [17]) it is known that such an $\mathcal{O}$-bimodule is isomorphic to $\mathcal{O}$ itself, with T_a corresponding to left multiplication L_a and S_a to right multiplication R_a. From this follows the uniqueness and the explicit description in the following proposition:

<u>There is a unique irreducible L-module of highest weight $\lambda_1 + \lambda_2$; the L_0-module structure of the highest weight-space W is the structure of L_0-module on O given by</u>

(12) $$w(D + H_1 \otimes a + H_2 \otimes b) = wD + aw + wb$$

for $w \in W(=O)$; $a,b \in O$; $D \in \mathrm{Der}(O)$.

Rather than carrying out the calculations verifying that O is indeed a $\lambda_1 + \lambda_2$-admissible L_0-module, we may be content with observing that there is an irreducible L-module of highest weight $\lambda_1 + \lambda_2$, namely the adjoint module L. Then the above analysis shows that the highest weight-space W is as claimed. Since L is central simple, the adjoint module is absolutely irreducible. Its dimension is 78.

5. REPRESENTATIONS OF $s\ell(3,O)$: $2\lambda_1 + \lambda_2$ AND $\lambda_1 + 2\lambda_2$.

For these weights λ, we shall display an irreducible λ-admissible L_0-module which, along with the associated irreducible L-module of highest weight λ, is absolutely irreducible. The information from the split case given in §1, combined with Proposition II.2, then shows that this module is unique. Let $\lambda = 2\lambda_1 + \lambda_2$; the other case is treated symmetrically, with explicit description of the L_0-module given below.

The underlying space W is again O; the action of L_0 on W is given by

(13) $$w(D + H_1 \otimes a + H_2 \otimes b) = wD + aw + w(a + \bar{b}).$$

Evidently H_1 and H_2 act as the scalars 2 and 1, respectively. To see that W is an L_0-module, note first that $\bar{b}D = (\tau(b)1 - b_0)D = -b_0D = -(bD) = \overline{bD}$, since D maps O into O_0. Thus if $E \in \mathrm{Der}(O)$, $w[D + H_1 \otimes a + H_2 \otimes b, E] = w[DE] + (aE)w + w(aE + \bar{b}E)$ $= wDE + (aw)E + (w(a + \bar{b}))E - [(wE)D + a(wE) + (wE)(a + \bar{b})]$.

Also it will be noted that if $T: O \to O$ is any linear transformation, then $[T, R_{\bar{b}}] = -[T, R_b]$. Thus, for example,

(14) $$(w(H_1 \otimes a))(H_2 \otimes b) - (w(H_2 \otimes b))(H_1 \otimes a)$$

$$= w[R_a + L_a, R_{\bar{b}}] = -w[R_a + L_a, R_b].$$

Now the case of $\lambda_1 + \lambda_2$ shows that (4)-(6) hold in O with $T = L$ and $S = R$. By (4) and (6), therefore,

$$-w[R_a + L_a, R_b] = -\frac{1}{3}wD_{a,b} - \frac{1}{3}wL_{[ab]} - \frac{2}{3}wR_{[ab]}.$$

In L, $[H_1 \otimes a, H_2 \otimes b] = -\frac{1}{3} D_{a,b} - \frac{1}{3} H_1 \otimes [ab] + \frac{1}{3} H_2 \otimes [ab]$, whose action on w is

$$-\frac{1}{3} wD_{a,b} - \frac{1}{3} [ab]w - \frac{1}{3} w[ab] + \frac{1}{3} w\overline{[ab]} = -\frac{1}{3} wD_{a,b} - \frac{1}{3} wL_{[ab]} - \frac{2}{3} wR_{[ab]},$$

agreeing with (14). The module-condition for pairs $H_1 \otimes a$, $H_1 \otimes b$ and for pairs $H_2 \otimes a$, $H_2 \otimes b$ is verified similarly. Evidently an L_0-submodule of W is an ideal in O, so W is an irreducible (even absolutely irreducible) L_0-module of weight $2\lambda_1 + \lambda_2$.

It remains only to show that W is $2\lambda_1 + \lambda_2$-admissible; that is, we must show that if $M_{a,b}$ and $N_{a,b}$ are the images in $\mathrm{End}_F(W)$ of the respective projections via ψ on $U(L_0)$ of $(E_{21} \otimes a)^3(E_{12} \otimes b)^3$ and $(E_{32} \otimes a)^2(E_{23} \otimes b)^2$ in $U(L)$, then $M_{a,b} = 0 = N_{a,b}$. Computation in L gives, for $\psi((E_{32} \otimes a)(E_{23} \otimes b))$ and for $\psi((E_{21} \otimes a)(E_{12} \otimes b))$, the respective values:

$$[E_{32} \otimes a, E_{23} \otimes b] = \frac{1}{3} D_{a,b} + \frac{2}{3} H_1 \otimes \frac{[ab]}{2} + H_2 \otimes \frac{ab+ba}{2} + \frac{1}{3} H_2 \otimes \frac{[ab]}{2}, \tag{15}$$

$$[E_{21} \otimes a, E_{12} \otimes b] = \frac{1}{3} D_{a,b} + H_1 \otimes \frac{ab+ba}{2} - \frac{1}{3} H_1 \otimes \frac{[ab]}{2} - \frac{2}{3} H_2 \otimes \frac{[ab]}{2}. \tag{16}$$

The respective actions of (15) and (16) and W are

$$\frac{1}{2}([S_a, S_b] + S_{ab+ba}), \tag{17}$$

$$\frac{1}{2}([T_a, T_b] + T_{ab+ba}). \tag{18}$$

From the definitions of S_a, T_a, we have for $w \in W$ ($= O$)

$$w\psi((E_{32} \otimes a)(E_{23} \otimes b)) = (\overline{wa})\overline{b} = wR_{\bar{a}} R_{\bar{b}}; \tag{19}$$

$$\begin{aligned} w\psi((E_{21} \otimes a)(E_{12} \otimes b)) &= (wa)b + b(aw) + (aw)b - a(wb) \\ &= w(R_a R_b + L_a L_b + [L_a, R_b]). \end{aligned} \tag{20}$$

Proceeding from (19) and (20), and using the rules for calculation in L, one finally derives:

$$N_{a,b} = 2(R_{\bar{a}} R_{\bar{b}})^2 - 2R_{\bar{a}\bar{b}\bar{a}} R_{\bar{b}}; \tag{21}$$

$$(22)\qquad M_{a,b} = 6(R_aR_b + L_aL_b + [L_a,R_b])^3$$
$$- 12(R_aR_b + L_aL_b + [L_a,R_b])(R_{aba}R_b + L_{aba}L_b + [L_{aba},R_b])$$
$$- 6(R_{aba}R_b + L_{aba}L_b + [L_{aba},R_b])(R_aR_b + L_aL_b + [L_a,R_b])$$
$$+ 12(R_{ababa}R_b + L_{ababa}L_b + [L_{ababa}, R_b]).$$

By the "Moufang identities" in alternative algebras (see [17], p. 16), $L_{aba} = L_aL_bL_a$ and $R_{aba} = R_aR_bR_a$. The second of these yields at once that $N_{a,b} = 0$.

To show $M_{a,b} = 0$ is more complicated. It was pointed out to the author by Kevin McCrimmon that its vanishing is a consequence of the theory of Jordan identities and the fact that $\mathcal{O}$ is a Jordan algebra with $a\cdot b = \frac{1}{2}(ab + ba)$ -- as is any alternative algebra with unit ([17], p. 15). From the alternative laws, one finds

$$w(R_aR_b + L_aL_b + [L_a,R_b]) = b(aw) + w(ab)$$
$$= (ba)w + (wa)b$$
$$= 2[(a\cdot b)\cdot w + (a\cdot w)\cdot b - a\cdot(w\cdot b)].$$

If we denote by $V_{a,b}$ (note the reversal of indices from [17]) the operator $w \to 2[(a\cdot b)\cdot w + (a\cdot w)\cdot b - a\cdot(w\cdot b)]$ in this Jordan algebra, then we have in $\mathcal{O}$,

$$R_aR_b + L_aL_b + [L_a, R_b] = V_{a,b}\ .$$

Also central to the Jordan theory is the mapping $U_a\colon \mathcal{O} \to \mathcal{O}$ sending w to $2(w\cdot a)\cdot a - w\cdot(a\cdot a)$. This last is equal to awa, since a and w generate an associative subalgebra. To show that $M_{a,b} = 0$ it will therefore be sufficient to show the identity,

$$(23)\qquad (V_{a,b})^3 - 2V_{a,b}V_{bU_a,b} - V_{bU_a,b}\,V_{a,b} + 2V_{aU_bU_a,b} = 0$$

in Jordan algebras.

When A is an <u>associative</u> algebra and $a\cdot b = \frac{1}{2}(ab + ba)$ makes A into a Jordan algebra, we have $wV_{a,b} = baw + wab$, $bU_a = aba$, and checking (23) is routine; one also finds $[V_{a,b}, V_{bU_a,b}] = 0$ in this case. Since our structure of Jordan algebra on A is a subalgebra of an associative algebra, <u>viz.</u> the Clifford algebra of the symmetric bilinear form $(a_0,b_0) = \frac{1}{2}(a_0\bar{b}_0 + b_0\bar{a}_0)$ on $\mathcal{O}_0$, the vanishing of (23) follows. (In fact, application of Macdonald's theorem ([17], p. 41) shows that the identity (23) holds in <u>all</u> Jordan algebras.)

The case of $\lambda_1 + 2\lambda_2$ is analogous, and we have proved

PROPOSITION IX.1. <u>There is a unique irreducible representation of</u> $L = s\ell(3,\mathcal{O})$ <u>with highest weight</u> $2\lambda_1 + \lambda_2$. <u>The highest weight-space identifies with</u> $\mathcal{O}$, <u>and the action of the general element</u>

$$g = D + H_1 \otimes a + H_2 \otimes b$$

<u>of</u> L_0 <u>on</u> $w \in \mathcal{O}$, <u>with</u>

$$wD + aw + w(a + \bar{b}). \tag{24}$$

<u>For the highest weight</u> $\lambda_1 + 2\lambda_2$, <u>the same conclusions hold, except that</u> (24) <u>is replaced by</u>

$$wD + (b + \bar{a})w + wb. \tag{25}$$

The corresponding irreducible representations of L are absolutely irreducible, and each has dimension 351. (They are contragredients of one another.)

<u>REMARKS ON TRIALITY:</u> As in §V.2.b of [26], $[L_0 L_0] = \mathrm{Der}(\mathcal{O}) + (H_1 \otimes \mathcal{O}_0) + (H_2 \otimes \mathcal{O}_0)$ is identified with $so(\mathcal{O})$, the Lie algebra of F-endomorphisms of $\mathcal{O}$ that are skew with respect to the norm form, by

$$D + (H_1 \otimes a_0) + (H_2 \otimes b_0) \leftrightarrow D + L_{a_0} + R_{b_0} .$$

The considerations of this section give two more such identifications:

$$D + (H_1 \otimes a_0) + (H_2 \otimes b_0) \leftrightarrow D + L_{a_0} + R_{a_0 - b_0} , \tag{26}$$

$$D + (H_1 \otimes a_0) + (H_2 \otimes b_0) \leftrightarrow D + L_{b_0 - a_0} + R_{b_0} . \tag{27}$$

There result automorphisms φ, ψ of $so(\mathcal{O})$, generating a group of order six, and incorporating the <u>principle of triality</u> for $so(\mathcal{O})$. Namely, if

$$(D + L_{a_0} + R_{b_0})^{\varphi} = D + L_{a_0} + R_{a_0 - b_0} \quad \text{and}$$

$$(D + L_{a_0} + R_{b_0})^{\psi} = D + L_{b_0 - a_0} + R_{b_0} ,$$

then $(xy)T = (xT^{\varphi})y + x(yT^{\psi})$ for all $x,y \in \mathcal{O}$, $T \in so(\mathcal{O})$. For $T = D$, the last is the definition of a derivation; for $T = L_{a_0}$ or R_{b_0}, it is immediate from the alternative law.

(The above realization of the "triality automorphisms" is indicated, although not so explicitly, in [18].)

6. REPRESENTATIONS OF $s\ell(3,\mathcal{O})$: $2\lambda_1 + 2\lambda_2$ AND SUMMARY.

By reference to Appendix 6, we see that there are exactly three irreducible representations in the split case whose highest weight restricts to $2\lambda_1 + 2\lambda_2$ on T. We propose to use this fact to show that all $2\lambda_1 + 2\lambda_2$-admissible irreducible L_0-modules occur in the tensor square of the irreducible $\lambda_1 + \lambda_2$-module $\mathcal{O}$ of §4. In this tensor square $\mathcal{O} \otimes \mathcal{O}$, we have

$$(u \otimes v)(H_1 \otimes a) = au \otimes v + u \otimes av,$$

$$(u \otimes v)(H_2 \otimes b) = ub \otimes v + u \otimes vb,$$

$$(u \otimes v)D = uD \otimes v + u \otimes vD.$$

The symmetric and skew tensors in $\mathcal{O} \otimes \mathcal{O}$ are evidently supplementary L_0-submodules, of respective dimensions 36 and 28. The bilinearized norm form (u,v) in $\mathcal{O}$ affords a mapping $\mathcal{O} \otimes \mathcal{O} \to F$ sending $u \otimes v$ to (u,v). If $a + \bar{a} = 0$ in $\mathcal{O}$, then $(au, v) + (u, av) = 0 = (ua, v) + (u, va)$, while $(uD, v) + (u, vD) = 0$ for each derivation D and $(1u, v) + (u, 1v) = 2(u, v) = (u1, v) + (v, v1)$. Thus if F is regarded as L_0-module annihilated by $[L_0\ L_0]$ and corresponding to the weight $2\lambda_1 + 2\lambda_2$ on the center T of L_0, the pairing $u \otimes v \to (u, v)$ defines a homomorphism of L_0-modules. The fact that $(u, v) = (v, u)$ means that the skew tensors in $\mathcal{O} \otimes \mathcal{O}$ lie in the kernel, as does a 35-dimensional submodule of the symmetric tensors, namely the set of those $\sum_i (u_i \otimes v_i + v_i \otimes u_i)$ with $\sum_i (u_i, v_i) = 0$.

The form (u, v) may be used in another way, namely to identify $\mathcal{O}$ with its dual $\mathcal{O}^*$ and thereby $\mathcal{O} \otimes \mathcal{O}$ with $\mathcal{O}^* \otimes \mathcal{O} = \mathrm{End}_F(\mathcal{O})$. The structure of L_0-module that results on $\mathrm{End}_F(\mathcal{O})$ is that where T acts by $2\lambda_1 + 2\lambda_2$ and $[L_0\ L_0]$ by the adjoint action on $\mathrm{End}_F(\mathcal{O})$ of the subalgebra $so(\mathcal{O})$. From this viewpoint, the 28-dimensional skew tensors form the Lie algebra $so(\mathcal{O})$, the submodule F is the scalar transformations, and the 35-dimensional submodule is the symmetric transformations of trace zero. As in §III.1 of [26], all three are absolutely irreducible modules for $[L_0\ L_0] = so(\mathcal{O})$, and consequently for L_0. By §II.2 they are $2\lambda_1 + 2\lambda_2$-admissible, and the resulting representations of L are absolutely irreducible, inequivalent over the algebraic closure, and have highest weights that restrict to $2\lambda_1 + 2\lambda_2$. By Proposition II.2, it follows that these three are a complete set of irreducible $2\lambda_1 + 2\lambda_2$-admissible L_0-modules.

In the sense of §V.7, we have now proved:

THEOREM IX.1. *A fundamental set for $s\ell(3,\mathcal{O})$ consists of the weights* $2\lambda_1$, $2\lambda_2$, $\lambda_1 + \lambda_2$, $2\lambda_1 + \lambda_2$, $\lambda_1 + 2\lambda_2$, *and for each of these weights* λ, *of a single irreducible* λ-*admissible* $\mathcal{L}_0$-*module* $\mathcal{W}^{(\lambda)}$. *For* $\lambda = 2\lambda_1, 2\lambda_2$, $\mathcal{W}^{(\lambda)}$ *is one-dimensional, as in* §3. *For* $\lambda = \lambda_1 + \lambda_2, 2\lambda_1 + \lambda_2, \lambda_1 + 2\lambda_2$, $\mathcal{W}^{(\lambda)}$ *is identified with* $\mathcal{O}$, *the structure of* $\mathcal{L}_0$-*module being given in* §§4.5. *All these fundamental modules are absolutely irreducible.*

7. AN EMBEDDING OF $\mathfrak{sl}(3,\mathcal{O})$ IN $sp(6,\mathcal{O})$.

In terms of the matrices describing $sp(6,F)$ in §2, we have an isomorphism of $\mathfrak{sl}(3,F)$ onto a subalgebra of $sp(6,F)$ sending $A \in \mathfrak{sl}(3,F)$ onto

$$\left(\begin{array}{c|c} A & 0 \\ \hline 0 & -A^s \end{array}\right),$$

where A^s is the transpose of JAJ, $J = \begin{pmatrix} 0 & 0 & 1 \\ 0 & 1 & 0 \\ 1 & 0 & 0 \end{pmatrix}$, so that $A \mapsto A^s$ is an involution in the matrix algebra $M_3(F)$. This map extends to a map of $\mathfrak{sl}(3,F) \otimes \mathcal{O}$ into $sp(6,\mathcal{O})$, sending $A \otimes b$, for $b \in \mathcal{O}$, to

$$\left(\begin{array}{c|c} A & 0 \\ \hline 0 & -A^s \end{array}\right) \otimes \tau(b) + \left(\begin{array}{c|c} A & 0 \\ \hline 0 & A^s \end{array}\right) \otimes b_0,$$

and this map, in turn, to a map φ of $\mathfrak{sl}(3,\mathcal{O}) = \mathrm{Der}(\mathcal{O}) \oplus (\mathfrak{sl}(3,F) \otimes \mathcal{O})$ into $sp(6,\mathcal{O})$ sending $D \in \mathrm{Der}(\mathcal{O}) \subseteq \mathfrak{sl}(3,\mathcal{O})$ to the same D in $\mathrm{Der}(\mathcal{O}) \subseteq sp(6,\mathcal{O})$.

Evidently the image of φ contains $\mathrm{Der}(\mathcal{O})$ and a 64-dimensional subspace of $sp(6,F) + (\mathcal{M}' \otimes \mathcal{O}_0) \subseteq sp(6,\mathcal{O})$, so that φ is a linear isomorphism of $\mathfrak{sl}(3,\mathcal{O})$ onto its image. In fact, φ *is a Lie morphism; thus the image is a subalgebra of* $sp(6,\mathcal{O})$ *isomorphic to* $\mathfrak{sl}(3,\mathcal{O})$. This observation will be used as was the corresponding one in §VII.3; here a direct proof seems simplest.

It suffices to test the effect of φ on brackets of the forms $[A \otimes a, D]$ and $[A \otimes a, B \otimes b]$. For the former, $[\varphi(A \otimes a), \varphi(D)]$

$$= \left[\left(\begin{array}{c|c} A & 0 \\ \hline 0 & -A^s \end{array}\right) \otimes \tau(a) + \left(\begin{array}{c|c} A & 0 \\ \hline 0 & A^s \end{array}\right) \otimes a_0, D\right]$$

$$= \left(\begin{array}{c|c} A & 0 \\ \hline 0 & A^s \end{array}\right) \otimes a_0 D = \left(\begin{array}{c|c} A & 0 \\ \hline 0 & A^s \end{array}\right) \otimes aD$$

$= \varphi(A \otimes aD) = \varphi([A \otimes a, D])$, since $aD \in \mathcal{O}_0$. For the latter,

$[\varphi(A \otimes a), \varphi(B \otimes b)] =$

$$\begin{pmatrix} [AB] & 0 \\ 0 & -[AB]^s \end{pmatrix} \otimes \tau(a)\,\tau(b) \;+\; \begin{pmatrix} [AB] & 0 \\ 0 & [AB]^s \end{pmatrix} \otimes \frac{\tau(a)b_0+\tau(b)a_0}{2}$$

$$+ \left[\begin{pmatrix} A & 0 \\ 0 & A^s \end{pmatrix} \otimes a_0,\; \begin{pmatrix} B & 0 \\ 0 & B^s \end{pmatrix} \otimes b_0 \right] .$$

The last bracket is equal to

$$\frac{1}{12}\,\mathrm{tr}(AB + A^sB^s)D_{a_0,b_0} \;+\; \begin{pmatrix} [AB] & 0 \\ 0 & -[AB]^s \end{pmatrix} \otimes \frac{a_0b_0+b_0a_0}{2}$$

$$+ \begin{pmatrix} AB+BA - \frac{2}{3}\,\mathrm{tr}(AB)I & 0 \\ 0 & A^sB^s+B^sA^s - \frac{2}{3}\,\mathrm{tr}(AB)I \end{pmatrix} \otimes \frac{[a_0b_0]}{2} .$$

On the other hand, in $\mathfrak{sl}(3,\mathcal{O})$, $[A \otimes a, B \otimes b]$

$$= \frac{1}{6}\,\mathrm{tr}(AB)D_{a,b} + [AB] \otimes \frac{ab+ba}{2} + (AB + BA - \frac{2}{3}\,\mathrm{tr}(AB)I) \otimes \frac{[ab]}{2} .$$

Now $[ab] = [a_0b_0]$ and $\mathrm{tr}(A^sB^s) = \mathrm{tr}(AB)$, so the first and third of these three terms have as image under φ the first and third in the expansion of the bracket above. Moreover, $\tau\left(\frac{ab+ba}{2}\right) = \tau(a)\tau(b) +$

$\frac{a_0b_0+b_0a_0}{2}$, $\left(\frac{ab+ba}{2}\right)_0 = \frac{\tau(a)b_0+\tau(b)a_0}{2}$, so that $\varphi\left([AB] \otimes \frac{ab+ba}{2}\right)$

is matched with the remaining terms in the expansion of $[\varphi(A \otimes a), \varphi(B \otimes b)]$.

The canonical split torus in $\mathfrak{sl}(3,\mathcal{O})$ spanned by $E_{22} - E_{11}$ and $E_{33} - E_{22}$ is mapped by φ onto a subspace of the maximal split torus $\mathcal{T}$ in $sp(6,\mathcal{O})$, with $\varphi(E_{22}- E_{11}) = H_1$, $\varphi(E_{33} - E_{22}) = H_2$. We are thus justified in using the same notation H_i, $i = 1,2$, for $E_{i+1,i+1} - E_{ii} \in \mathfrak{sl}(3,\mathcal{O})$ and for its image under φ. The corresponding centralizer $\mathfrak{sl}(3,\mathcal{O})_0 = \mathrm{Der}(\mathcal{O}) + H_1 \otimes \mathcal{O} + H_2 \otimes \mathcal{O}$ is sent by φ into the centralizer of $\mathcal{T}$ in $sp(6,\mathcal{O})$, since $\varphi(H_i \otimes a) = \tau(a)H_i + U_i \otimes a_0$, $i = 1,2$. From this we also see that

$$L_0 = \varphi(\mathfrak{sl}(3,\mathcal{O})_0) + FH_3 .$$

The images under φ of the spaces designated $L_{\pm\alpha_1}, L_{\pm\alpha_1}$ in $\mathfrak{sl}(3,\mathcal{O})$ are the spaces given the same designations in $sp(6,\mathcal{O})$ (see §§1,2), and corresponding root-spaces will be identified. We are now in the setting arrived at in §VIII.3, with the root-spaces $L_{\pm\alpha_3}$ in $sp(6,\mathcal{O})$ again having dimension one. If $\lambda \in T^*$ is dominant integral, and if $\mathcal{W}$ is an irreducible λ-admissible L_0-module, then $\mathcal{W}$ is an irreducible $(\lambda \circ \varphi)$-admissible $s\ell(3,\mathcal{O})_0$-module, and conversely every irreducible $(\lambda \circ \varphi)$-admissible irreducible $s\ell(3,\mathcal{O})_0$-module becomes an irreducible λ-admissible L_0-module by letting H_3 act by $\lambda(H_3)$. We apply this fact and our theory for $s\ell(3,\mathcal{O})$ in the next section.

8. FUNDAMENTAL REPRESENTATIONS FOR $sp(6,\mathcal{O})$.

We enumerate all λ-admissible irreducible L_0-modules for the seven values of λ listed in §2.

$\lambda = \lambda_3$: Here λ vanishes on H_1 and H_2. As in §3, any irreducible λ-admissible L_0-module $\mathcal{W}$, considered as module for $s\ell(3,\mathcal{O})_0$, must be trivial. Thus $\mathcal{W}$ is one-dimensional with the general element

$$g = D + \sum_{i=1}^{3} \mu_i H_i + (U_1 \otimes a_0) + (U_2 \otimes b_0)$$

of L_0 sending $w \in \mathcal{W}$ to $\mu_3 w$. The corresponding representation of L is absolutely irreducible; upon extending to a splitting field, it is the representation of highest weight π_1 (see Appendix 6), of dimension 56.

$\lambda = 2\lambda_1$: The unique irreducible λ-admissible L_0-module $\mathcal{W}$ is the irreducible $2\lambda_1$-admissible $s\ell(3,\mathcal{O})_0$-module, with

$$wg = 2\mu_1 w, \quad g \text{ as above.}$$

Again the representation of L is absolutely irreducible; it is the adjoint representation, of dimension 133.

$\lambda = 2\lambda_2$: As for $2\lambda_1$, with

$$wg = 2\mu_2 w.$$

The corresponding representation of $\mathcal{W}$ is absolutely irreducible, of highest weight π_2 over a splitting field, and is seen from Weyl's formula to have dimension 1539.

$\lambda = \lambda_1 + \lambda_2$: From §3 and the reduction to $s\ell(3,\mathcal{O})$, we see that the unique $\lambda_1 + \lambda_2$-admissible irreducible L_0-module $\mathcal{W}$ identifies with $\mathcal{O}$, with

$$wg = wD + (\mu_1 + \mu_2)w + a_0 w + wb_0.$$

The absolute irreducibility of $\mathcal{W}$ implies that of the corresponding L-module. Over a splitting field, the latter has highest weight π_7. Its dimension is 912.

$\lambda = 2\lambda_1 + \lambda_2$: Again $\mathcal{W} = \mathcal{O}$, now with

$$wg = wD + (2\mu_1 + \mu_2)w + a_0 w + w(a_0 - b_0),$$

as in §5. Once more $\mathcal{W}$ is absolutely irreducible and the representation of L is an F-form of the irreducible representation of highest weight π_5 in the split case. The L-module has dimension 8465.

$\lambda = \lambda_1 + 2\lambda_2$: The situation is as in the last case,

$$wg = wD + (\mu_1 + 2\mu_2)w + (b_0 - a_0)w + wb_0.$$

The corresponding irreducible L-module is still absolutely irreducible. Upon splitting, it has highest weight π_3; its dimension is 27,664.

$\lambda = 2\lambda_1 + 2\lambda_2$: The three choices for $\mathcal{W}$ here are obtained by decomposing the L_0-module $\mathcal{O} \otimes \mathcal{O}$, where $\mathcal{O} = \mathcal{W}$ for $\lambda_1 + \lambda_2$ above, into:

i) a one-dimensional module;

ii) the skew tensors, of dimension 28;

iii) the symmetric tensors "of trace zero", as in the corresponding part of §6, a module of dimension 35.

All three of these L_0-modules are absolutely irreducible, from which it follows that the corresponding L-modules are likewise. Over a splitting field, the highest weights are, respectively:

i) $\pi_2 + \pi_6$; ii) π_4; iii) $2\pi_7$.

THEOREM IX.2. A fundamental set for $sp(6,\mathcal{O})$ consists of the six weights $2\lambda_1, 2\lambda_2, \lambda_3, \lambda_1 + \lambda_2, 2\lambda_1 + \lambda_2, \lambda_1 + 2\lambda_2$ and, for each of these weights λ, of a unique λ-admissible irreducible L_0-module $\mathcal{W}^{(\lambda)}$. This module is given explicitly in the list above.

9. FUNDAMENTAL REPRESENTATIONS FOR $F_4(\mathcal{O})$.

The key here is to note that, as in §7 above for $s\ell(3,\mathcal{O})$ and $sp(6,\mathcal{O})$, and as in §VIII.4 for $s\ell(3,\mathcal{Q})$ in $F_4(\mathcal{Q})$, we have an embedding of $s\ell(3,\mathcal{O})$ in $F_4(\mathcal{O})$. The first part of §VIII.4 was carried out assuming only that $\mathcal{D}$ is a composition division algebra, allowing $\mathcal{D} = \mathcal{O}$ as a possibility. The map there called θ_0 becomes here a map $\varphi_0 : s\ell(3,\mathcal{O})_0 \to M_0$ with exactly the same definition as on p. 114 with $\mathcal{Q}$ replaced by $\mathcal{O}$, and is again an isomorphism of Lie algebras. The choice of representatives $\omega(w)$, $\omega'(w)$ for elements of the Weyl groups is made as before, and, as there, each $\varphi_0 \omega' \varphi_0^{-1} \omega^{-1}$ is the identity on $s\ell(3,\mathcal{O})_0$. The isomorphism of the $s\ell(3,\mathcal{O})_0$-module $E_{23} \otimes \mathcal{O}$ onto $L_{\alpha_3} \subseteq M$ is defined and checked as in §VIII.3, and Allison's theorem applies to yield an isomorphism φ of $s\ell(3,\mathcal{O})$

onto M extending φ_0 and mapping root-spaces as in §VIII.3, with $\mathcal{Q}$ replaced by $\mathcal{O}$. Again one has $L_0 = \varphi(\mathfrak{sl}(3,\mathcal{O})_0) + FH_1 + FH_2$, and the theory of λ-admissible irreducible L_0-modules is essentially the same as that of $\lambda \circ \varphi_0$ -admissible irreducible $\mathfrak{sl}(3,\mathcal{O})_0$-modules.

From §VIII.1, the weights λ on T which must be considered are $\lambda_1, \lambda_2, 2\lambda_3, 2\lambda_4, \lambda_3 + \lambda_4, 2\lambda_3 + \lambda_4, \lambda_3 + 2\lambda_4, 2\lambda_3 + 2\lambda_4$. Now the conclusions run parallel to those of §8:

$\lambda = \lambda_1$: Here $\lambda \circ \varphi_0 = 0$; any irreducible λ-admissible L_0-module W is a trivial $\mathfrak{sl}(3,\mathcal{O})_0$-module, so is one-dimensional, with the general element

$$g = D + \sum_{i=1}^{4} \mu_i H_i + (U_3 \otimes a_0) + (U_4 \otimes b_0)$$

of L_0 sending $w \in W$ to $\mu_1 w$. The representation of L is absolutely irreducible, and is the adjoint representation, of dimension 248.

$\lambda = \lambda_2$: The situation is as for λ_1, with $\mu_1 w$ replaced by $\mu_2 w$. Again the representation of L is absolutely irreducible, and becomes upon splitting the representation with highest weight π_2, in the notation of Appendix 5. Its dimension is 30,380.

$\lambda = 2\lambda_4$: The correspondence with representations of $\mathfrak{sl}(3,\mathcal{O})$ shows that here the unique irreducible $2\lambda_4$-admissible L_0-module W has dimension one, with $wg = 2\mu_4 w$. The corresponding irreducible L-module is absolutely irreducible; upon splitting, it has highest weight π_7, and its dimension is 3875.

$\lambda = 2\lambda_3$: The situation is as for $2\lambda_4$, with $2\mu_3 w$ for $2\mu_4 w$. The corresponding irreducible L-module is absolutely irreducible, with highest weight π_3 upon splitting; its dimension is 2,450,240.

$\lambda = \lambda_3 + \lambda_4$: Here the highest weight-space W of the unique irredudible L-module with this highest weight is $W = \mathcal{O}$, with

$$wg = wD + (\mu_3 + \mu_4)w - b_0 w + wa_0,$$

by our map of $\mathfrak{sl}(3,\mathcal{O})$ into L. The absolute irreducibility of W implies that the L-module is absolutely irreducible; upon splitting, it becomes the module of highest weight π_8; its dimension is 147,250.

$\lambda = \lambda_3 + 2\lambda_4$: Again $W = \mathcal{O}$, now with

$$wg = wD + (\mu_3 + 2\mu_4)w - b_0 w - w(a_0 + b_0).$$

The corresponding L-module is again absolutely irreducible, having highest weight π_6 in the split case, and dimension 6,696,000.

$\lambda = 2\lambda_3 + \lambda_4$: Here $W = \mathcal{O}$, with

$$wg = wD + (2\mu_3 + \mu_4)w + (a_0 + b_0)w + wa_0.$$

The corresponding irreducible L-module is absolutely irreducible, with highest weight π_4 in the split case, and dimension 146,325,270.

$\lambda = 2\lambda_3 + 2\lambda_4$: Again one decomposes the tensor square $\mathcal{O} \otimes \mathcal{O}$ of the $\lambda_3 + \lambda_4$-admissible L_0-module $\mathcal{O}$ into irreducible submodules of dimensions 1, 28, 35. All three resulting irreducible L-modules of highest weight $2\lambda_3 + 2\lambda_4$ are absolutely irreducible, becoming upon splitting those of highest weights $\pi_3 + \pi_7$; π_5; and $2\pi_8$. By Proposition II.2, the three irreducible λ-admissible L_0-modules are exhaustive.

We summarize our results on $F_4(\mathcal{O})$ in

THEOREM IX.3. *A fundamental set for* $F_4(\mathcal{O})$ *consists of the seven weights* $\lambda_1, \lambda_2, 2\lambda_3, 2\lambda_4, \lambda_3 + \lambda_4, 2\lambda_3 + \lambda_4, \lambda_3 + 2\lambda_4$ *and, for each of these weights* λ, *of a unique* λ-*admissible irreducible* L_0-*module* $W^{(\lambda)}$. *This module is explicit in the list above.*

CHAPTER X

EXCEPTIONAL TYPES III: RELATIVE TYPE A_1

The algebras of the title are those obtained by the construction of Koecher and Tits, based on a central Jordan division algebra J, as in §III.2 of [26]. That is, the underlying vector space of L is

$$\mathrm{Der}(J) \oplus (\mathfrak{sl}(2,F) \otimes J),$$

with composition discussed below. The classification of simple Jordan algebras in [17] leaves only the following possibilities for J:

i) a central associative division algebra;

ii) the fixed elements of a central associative division algebra with respect to an involution of first kind;

iii) the fixed elements of a central associative division algebra over a quadratic extension of F, with respect to an involution of second kind;

iv) the Jordan algebra of a quadratic form in 2 or more dimensions, not representing 1;

v) an exceptional algebra, of dimension 27.

From §V.2 of [26], we see that i) is just the case $\mathfrak{sl}(2,\mathcal{D})$, treated in Chapter V, and from §V.3.d of [26], that the case iv) is that of Chapter VII (for relative rank one). In §V.7.a of [26], the cases ii) and iii) are likewise identified with Lie algebras of rank one treated in Chapters V resp. VI.

In this chapter, we treat the remaining case, that where J is an exceptional central Jordan division algebra. In preparation for the next chapter, we also develop some information about the corresponding Lie algebra when J is a cubic extension field of F. In the latter case, our algebra $\mathfrak{sl}(2,J)$ is not central simple. From Proposition I.3, this circumstance is no trouble.

1. THE LIE ALGEBRAS $\mathfrak{sl}(2,J)$ AND THEIR FUNDAMENTAL WEIGHTS.

The exceptional central Jordan division algebra J has dimension 27, unit element 1 and a linear trace function $t\colon J \to F$ with $t(1) = 3$ and, for each $a \in J$,

$$a^3 - t(a)a^2 + \frac{1}{2}(t(a)^2 - t(a^2))a - \frac{1}{6}(t(a)^3 - 3t(a)t(a^2) + 2t(a^3))1 = 0.$$

If a and 1 are linearly independent in J, then the polynomial in $F[X]$,

$$p_a(X) = X^3 - t(a)X^2 + \frac{1}{2}(t(a)^2 - t(a^2))X - \frac{1}{6}(t(a)^3 - 3t(a)t(a^2) + 2t(a^3)),$$

is irreducible and is the <u>generic minimum polynomial</u> of a.

Upon passage to an extension field K, J_K becomes isomorphic to the "split exceptional Jordan algebra over K", as encountered in Chapters VIII and IX in connection with algebras of type F_4. The kernel J_0 of t is stable under the action of $\mathrm{Der}(J)$ and under the bilinear product $x \circ y = xy - \frac{1}{3}t(xy)1$. (Here xy denotes the commutative product natural to J.)

Our Lie algebra $L = \mathfrak{sl}(2,J)$ is defined as the vector space

$$L = \mathrm{Der}(J) \oplus (\mathfrak{sl}(2,F) \otimes J),$$

with compositions as follows:

$\mathrm{Der}(J)$ is a subalgebra, with the usual bracket;

For $D \in \mathrm{Der}(J)$, $x \in \mathfrak{sl}(2,F)$, $a \in J$,

$$[x \otimes a, D] = x \otimes aD;$$

For $x,y \in \mathfrak{sl}(2,F)$ and $a,b \in J$,

$$[x \otimes a, y \otimes b] = 2t(xy)D_{a,b} + [xy] \otimes ab,$$

where $D_{a,b}$ is the inner derivation of J sending $c \in J$ to $(ca)b - (cb)a$.

A maximal split torus T in L is spanned by $H \otimes 1$, where $H = E_{22} - E_{11} \in \mathfrak{sl}(2,F)$. The centralizer of T is

$$L_0 = \mathrm{Der}(J) + (H \otimes J),$$

and $L = L_0 + L_{\alpha_1} + L_{-\alpha_1}$, where $L_{\alpha_1} = E_{12} \otimes J$, $L_{-\alpha_1} = E_{21} \otimes J$, is the decomposition of L into root-spaces relative to T. Let $\lambda_1 \in T^*$, $\lambda_1(H_1) = 1$; from the information on splitting to be found in Appendix 7, the restrictions to T of the highest weights of the fundamental representations in the split case are: $2\lambda_1$, $3\lambda_1$, $4\lambda_1$, $5\lambda_1$ and $6\lambda_1$. The respective numbers of irreducible representations of the split algebra whose highest weights have these restrictions are: 1, 2, 3, 3, 7.

2. <u>FUNDAMENTAL REPRESENTATIONS FOR $\mathfrak{sl}(2,J)$: FIRST IDENTITIES.</u>

Let k be a positive integer (soon only $k = 2, 3$ will be considered), and let W be the highest weight-space for an irreducible $\mathfrak{sl}(2,J)$-module of highest weight $k\lambda_1$. Let $\rho(a)$ be the linear transformation $w \to w(H \otimes a)$ of W, so that $\rho(1) = k1$. The condition of $k\lambda_1$-admissibility means that, for all a and $b \in J$, all $w \in W$,

$$w\psi((E_{21}\otimes a)^{k+1}(E_{12}\otimes b)^{k+1}) = 0.$$

Now we have

$$w\psi((E_{21}\otimes a)(E_{12}\otimes 1)) = w[E_{21}\otimes a, E_{12}\otimes 1]$$
$$= w(H\otimes a) = w\rho(a), \quad \text{and}$$
$$w\psi((E_{21}\otimes a)^2(E_{12}\otimes 1)^2) = w\rho(a)^2$$
$$+ w\rho(a)\psi((E_{21}\otimes a)(E_{12}\otimes 1)) - 2w\psi((E_{21}\otimes a^2)(E_{12}\otimes 1))$$
$$= 2w(\rho(a)^2 - \rho(a^2)).$$

Thus if $k = 0$, $\rho(a) = 0$ for all a while if $k = 1$, $\rho(a^2) = \rho(a)^2$ for all a. The latter implies $\rho(ab) = \frac{1}{2}(\rho(a)\rho(b) + \rho(b)\rho(a))$, or that W is a special J-module with the representation ρ. But the fact that J is not special again implies $\rho = 0$, contradicting $\rho(1) = 1$. Thus $k = 1$ is impossible, offering a "rational" reason why there is no irreducible representation of highest weight λ_1.

We propose to exhibit a $2\lambda_1$-admissible irreducible L_0-module, and two inequivalent $3\lambda_1$-admissible modules, having inequivalent submodules upon splitting. The irreducible L-modules with these as highest weight-spaces are absolutely irreducible. From the numbers in the split case determined at the end of §1 it will follow that these are all the irreducible L-modules of highest weights $2\lambda_1$ and $3\lambda_1$.

First for $\lambda = 2\lambda_1$, define an action (on the right) of L_0 on J by

$$b(D + H\otimes a) = bD + 2ab$$

for $b \in J$, $D \in \mathrm{Der}(J)$, $a \in J$. Because $[H\otimes a_1, H\otimes a_2] = 4D_{a_1,a_2}$, it is immediate that J is an irreducible (indeed absolutely irreducible) L_0-module, with $b\rho(a) = 2ba$, such that $\rho(1) = 2I$. That is, $\rho(a): J \to J$ is the mapping $2R_a$, R_a denoting mutliplication by a.

The linear isomorphism $b \mapsto E_{12}\otimes b$ from J to $L_{\alpha_1} \subseteq L$ is an isomorphism of L_0-modules; indeed, our definition of L_0-module structure on J was chosen to make this the case. We have $\alpha_1 = 2\lambda_1$ and L_{α_1} is the highest weight-space of the (absolutely irreducible) adjoint module for L. Thus L_{α_1}, or J with the above structure, is the unique $2\lambda_1$-admissible irreducible L_0-module, and is absolutely irreducible. The necessary condition for $\lambda(= 2\lambda_1)$-admissibility then gives certain identities concerning J, identities we investigate here and in later sections.

As above the necessary condition for λ-admissibility, "$w\psi((E_{21}\otimes a)^3(E_{12}\otimes 1)^3) = 0$" amounts to $6\rho(a)^3 - 18\rho(a)\rho(a^2) + 12\rho(a^3) = 0$, or to

$$(1) \qquad R_{a^3} - 3R_a R_{a^2} + 2R_a^3 = 0.$$

The identity (1) holds in all Jordan algebras A over F; for if A is special, so that the Jordan product $a \cdot b$ is $\frac{1}{2}(ab + ba)$, in a subspace of an associative algebra closed under this product (juxtaposition "ab" now denoting the associative product), (1) is easily verified. In particular, (1) holds in the free Jordan algebra on two generators, because that algebra is special by a theorem of Shirshov ([17], p. 47). It follows that (1) holds in all Jordan algebras.

We shall later use another device for proving identities in $\mathcal{J}$. To illustrate how this device yields (1), note first that it suffices to prove (1) for $\mathcal{J}_K$, where K is an algebraic closure of F. Thus we may assume $\mathcal{J}_K$ is the split exceptional Jordan algebra studied in connection with algebras $F_4(\mathcal{D})$ in Chapters VIII and IX. An argument based on density in the Zariski topology as in [17] (p. 14) then shows that we may assume the generic minimum polynomial of a, $p_a(X)$, has three distinct non-zero roots, so that $K[a] \subseteq \mathcal{J}_K$ has the form $K[a] = Ke_1 + Ke_2 + Ke_3$, where e_1, e_2, e_3 are orthogonal non-zero idempotents. As in [17], pp. 361-2, there is an automorphism of $\mathcal{J}_K$ mapping $K[a]$ onto the diagonal matrices in the usual realization of $\mathcal{J}_K$ as 3 by 3 hermitian matrices over $\mathcal{O}_K$. Thus a may be assumed to be such a diagonal matrix, and in that case it is an easy matter to verify (1), applying the operator on the left-hand side to a general $b \in \mathcal{J}_K$.

Now pass on to $\lambda = 3\lambda_1$, where we offer two candidates for $\mathcal{W}$:

a) $\mathcal{W} = F$, with $w(D + (H \otimes a)) = t(a)w$;

b) $\mathcal{W} = (\mathrm{Der}\ \mathcal{J}) \times \mathcal{J}_0$, with

$$(E, b_0)(D + H \otimes a) =$$

$$([E,D] + t(a)E + 4D_{b_0,a}, b_0D - aE + t(a)b_0),$$

for $E \in \mathrm{Der}(\mathcal{J})$, $b_0 \in \mathcal{J}_0$.

In a) it is clear that $\mathcal{W}$ is an absolutely irreducible $\mathcal{L}_0$-module, on which $H = H \otimes 1$ acts as multiplication by 3. The "first identities" of the title of this section refer to the case $b = 1$ of the general admissibility condition

$$w\psi((E_{21} \otimes a)^{k+1}(E_{12} \otimes b)^{k+1}) = 0,$$

for $k = 2,3$. Thus in a), b) above, we consider this relation for $k = 3$ and $b = 1$. By computation within $\mathcal{L}$ (or $\mathcal{U}(\mathcal{L})$), the meaning of this "first identity" in case a) is that

(2) $$t(a)^4 - 6t(a)^2\, t(a^2) + 3t(a^2)^2 + 8t(a)\, t(a^3) - 6t(a^4) = 0$$

for all $a \in J$.

That (2) holds is an immediate consequence of the fundamental cubic identity $p_a(a) = 0$ in J, since the left-hand side of (2) is $-6t(ap_a(a))$.

Next consider W as in b). Comparison with the definition of operations in L shows that

$$(E, b_0) \to E + (H \otimes b_0)$$

is a linear isomorphism of W onto an L_0-submodule of the adjoint module for L_0, indeed onto $[L_0L_0]$. Here the Lie algebra $[L_0L_0]$ is a central simple Lie algebra (without nilpotent elements) of dimension 78 and of split type E_6 (see [13],[26]), and thus W is an absolutely irreducible L_0-module. Again $H \otimes 1$ acts by multiplication by $t(1) = 3$. The "first identity" $w\psi((E_{21} \otimes a)^4(E_{12} \otimes 1)^4) = 0$ reduces to

(3) $$\rho(a)^4 - 6\rho(a)^2\rho(a^2) + 3\rho(a^2)^2 + 8\rho(a)\, \rho(a^3) - 6\rho(a^4) = 0,$$

for all $a \in J$. (Along the way, one uses that $\rho(a)$ and $\rho(a^2)$ commute; this follows from $[H \otimes a, H \otimes a^2] = 4D_{a,a^2} = 0$, the last by the fundamental Jordan identity $(ab)a^2 = (a^2b)a$. Then $\rho(a)$ and $\rho(a^3)$ commute because $p_a(a) = 0$ shows that a^3 is a linear combination of $1, a, a^2$.)

By our general remarks concerning identities, it suffices to verify (3) in the algebraically closed case, and there for elements a such that $p_a(X)$ has three distinct non-zero roots. Next, if φ is an automorphism of J, then

$$(E, b_0)\, \rho(a^\varphi) = (t(a^\varphi)E + 4D_{b_0,a^\varphi}, - a^\varphi E + t(a^\varphi)b_0) .$$

From $t(a^\varphi) = t(a)$ and $\varphi^{-1}D_{b,a}\varphi = D_{b^\varphi,a^\varphi}$, we see that defining

$(E,b_0)^\varphi = (\varphi^{-1}E\varphi, b_0^\varphi)$ gives a linear automorphism φ of W such that $w^\varphi\rho(a^\varphi) = (w\rho(a))^\varphi$ for all $w \in W$, $a \in J$. Thus if (3) holds for a, it holds for a^φ. By the second principle cited in our general attack on identities, we now may assume not only that J is split and canonically identified with 3 by 3 hermitian matrices over the split octonions O, but also that a is a diagonal matrix.

Now we verify (3) directly for the action of $\rho(a)$ on an element $(E,0)$ and on an element $(0, b_0)$. As in [24], we represent each of these canonically, writing for

$$b_0 = \begin{pmatrix} \beta_1 & u & v \\ \bar{u} & \beta_2 & w \\ \bar{v} & \bar{w} & \beta_3 \end{pmatrix} .$$

with $u, v, w \in \mathcal{O}$; $\beta_i \in F$, $\Sigma\beta_i = 0$, that $b_0 = \text{diag}\ \{\beta_1, \beta_2, \beta_3\} + u(1,2) + v(1,3) + w(2,3)$, and representing the derivation E as $E = (u, v, w, T)$, where $(\text{diag}\ \{\alpha_1, \alpha_2, \alpha_3\})E = \frac{1}{2}[(\alpha_1 - \alpha_2)u(1,2) + (\alpha_1 - \alpha_3)v(1,3) + (\alpha_2 - \alpha_3)w(2,3)]$ and where $x(1,2)E$ has $\frac{1}{2}xT$ in the 1,2-position. Here $u, v, w, x \in \mathcal{O}$ and T is a skew linear transformation of $\mathcal{O}$ with respect to the norm form (u,v).

Let $a = \text{diag}\ \{\alpha_1, \alpha_2, \alpha_3\}$. If $E = (0, 0, 0, T)$, then $aE = 0$ (see [24]), so that $\rho(a)$ sends $(E,0)$ to $t(a)(E,0)$. In this case, (3) amounts to (2) which has already been verified. Likewise, if b_0 is diagonal, we have $D_{b_0,a} = 0$ and the effect of $\rho(a)$ on $(0, b_0)$ is $t(a)\ (0, b_0)$, and again (3) holds on $(0, b_0)$. Next consider simultaneously the effect of $\rho(a)$ on $((u, 0, 0, 0), 0)$ and on $(0, u(1,2))$, for $u \in \mathcal{O}$: We find

$$((u, 0, 0, 0), 0)\ \rho(a) = (t(a)(u, 0, 0, 0), \tfrac{1}{2}(\alpha_2 - \alpha_1)\ u(1,2));$$

$$(0, u(1,2))\ \rho(a) = (2(\alpha_2 - \alpha_1)(u,0,0,0), t(a)\ u(1,2)).$$

It is a routine matter to verify that the left-hand side of (3) annihilates $(0, u(1,2))$ and $((u, 0, 0, 0), 0)$. The same procedure applies to $((0, u, 0, 0), 0)$ and $(0, u(1,3))$, as well as to $((0, 0, u, 0), 0)$ and $(0, u(2,3))$. The universal validity of the first identity (3) follows.

3. SECOND IDENTITIES.

We write simply α for the root we have previously denoted by α_1. If $x \in L_\alpha \cup L_{-\alpha}$, then $(\text{ad}\ x)^3 = 0$, so that $\exp(\text{ad}\ x)$ is a well-defined automorphism of L. In particular, if b is an invertible element of J, then $\varphi_b = \exp(\text{ad}(E_{12} \otimes b^{-1}))\exp(\text{ad}(-E_{21} \otimes b))\exp(\text{ad}(E_{12} \otimes b^{-1}))$. $\exp(\text{ad}(E_{12} \otimes 1))\exp(\text{ad}(-E_{21} \otimes 1))\exp(\text{ad}(E_{12} \otimes 1))$ is an automorphism fixing $H = H \otimes 1$, thus mapping L_0 to L_0 and stabilizing each of L_α, $L_{-\alpha}$. Using the fact that $D_{b,1} = D_{b^{-1},1} = D_{b,b^{-1}} = 0$, we calculate the effect of φ_b on $E_{12} \otimes 1$:

$$E_{12} \otimes 1 \xrightarrow{\varphi_b} E_{12} \otimes b^2.$$

The automorphism φ_b extends to a unique automorphism, also denoted φ_b, of the universal algebra $U(L)$, stabilizing each of the summands in the decomposition

$$U(L) = U(L_0) \oplus NU(L_0 + N) \oplus U(L_0 + N^-)N^- \oplus NU(L)N^-$$

Where $N = L_\alpha$, $N^- = L_{-\alpha}$. Thus ψ, the usual projection on $U(L_0)$ associated with this decomposition, commutes with φ_b. It follows that the restriction of φ_b to $U(L_0)$ maps

$$\psi(x^j(E_{12} \otimes 1)^j) \quad (x \in L_{-\alpha}) \quad \text{to}$$

(4) $$\psi((x^{\varphi_b})^j (E_{12} \otimes b^2)^j).$$

For $D \in \mathrm{Der}(J) \subseteq L_0$, we find

(5) $$D^{\varphi_b} = (D + 2D_{b^{-1},bD}) \oplus (H \otimes b(b^{-1}D)),$$

while for $a \in J$,

(6) $$(H \otimes a)^{\varphi_b} = 4D_{ab^{-1},b} \oplus (H \otimes (-a + 2(ab^{-1})b)).$$

For each of the L_0-modules a), b) of weight $3\lambda_1$ of §2 we define below a linear automorphism γ_b of the underlying vector space W such that for all $w \in W$, $x \in L_0$,

(7) $$w\gamma_b^{-1}\, x\gamma_b = w(x^{\varphi_b}) .$$

(It is also possible to define such a γ_b for the $2\lambda_1$-admissible module J, namely $\gamma_b = U_b$, in the notation of Jacobson: $a \to 2(ab)b - ab^2$; we shall not need this, because $2\lambda_1$-admissibility is known already.)

In case a), where $W = F$, we define γ_b to be the identity mapping; in case b), where $W = \mathrm{Der}(J) \times J_0$, we define

(8) $$\gamma_b = \rho(b)^3 - 3\rho(b)\,\rho(b^2) + 2\rho(b^3),$$

where $\rho(b)$ is defined in §2.

Once the relation (7) is established we shall have, for all $w \in W$, $x = x_\alpha \in L_\alpha$,

$$w\gamma_b\, \psi(x^4(E_{12} \otimes 1)^4)^{\varphi_b} = w\gamma_b\, \psi((x^{\varphi_b})^4(E_{12} \otimes b^2)^4)$$
$$= w\psi\,(x^4(E_{12} \otimes 1)^4)\gamma_b.$$

The "first identities" of §2 show that the last of these is zero for all w, all x_α, all b, so that the second is also zero. For fixed invertible b, $x_\alpha^{\varphi_b}$ runs over L_α as x_α does, and $w\gamma_b$ runs over W as w does. Thus the identical vanishing of the second member tells us that

$$\psi((E_{21} \otimes a)^4 (E_{12} \otimes b^2)^4)$$

<u>annihilates</u> W <u>for all invertible</u> $b \in J$, <u>all</u> $a \in J$. The vanishing of these quantities is the set of "second identities" referred to in the heading of this section. (Again there is a corresponding line of reasoning with $W = J$, $\gamma_b = U_b$, and with "4" replaced by "3", but it is unnecessary for our purposes.)

To verify (7), we may assume that J is split and that $p_b(X)$ has distinct roots, where $b \in J$ is invertible. As in §2, an automorphism of J may be applied, so that we may assume $b = \mathrm{diag}\ \{\beta_1, \beta_2, \beta_3\}$ in the usual realization of the split J, with $b^{-1} = \mathrm{diag}\ \{\beta_1^{-1}, \beta_2^{-1}, \beta_3^{-1}\}$. In the notations of the last section, we see from (5) that

$$(0, 0, 0, T)^{\varphi_b} = (0, 0, 0, T);$$

$$\begin{aligned}(u, 0, 0, 0)^{\varphi_b} &= ((u, 0, 0, 0) + D_{b^{-1}, (\beta_1 - \beta_2)u(1,2)} \\ &\quad + (H \otimes \tfrac{1}{2} (\beta_1^{-1} - \beta_2^{-1}) b \cdot u(1,2)) \\ &= \tfrac{1}{2} (\beta_1 \beta_2^{-1} + \beta_2 \beta_1^{-1}) (u, 0, 0, 0) \\ &\quad + \tfrac{1}{4} H \otimes (\beta_2 \beta_1^{-1} - \beta_1 \beta_2^{-1}) u(1,2).\end{aligned}$$

Then $(0, u, 0, 0)^{\varphi_b}$ and $(0, 0, u, 0)^{\varphi_b}$ are analogous to $(u, 0, 0, 0)^{\varphi_b}$; one replaces $(u, 0, 0, 0)$ by $(0, u, 0, 0)$, $u(1,2)$ by $u(1,3)$ and β_2 by β_3 in the former case and $(u, 0, 0, 0)$ by $(0, 0, u, 0)$, $u(1,2)$ by $u(2,3)$, β_1 by β_2 and β_2 by β_3, in the latter. Also

$$(H \otimes \mathrm{diag}\ \{\alpha_1, \alpha_2, \alpha_3\})^{\varphi_b} = H \otimes \mathrm{diag}\ \{\alpha_1, \alpha_2, \alpha_3\};$$

$$\begin{aligned}(H \otimes u(1,2))^{\varphi_b} &= (\beta_2 \beta_1^{-1} - \beta_1 \beta_2^{-1}) (u, 0, 0, 0) + \\ &\quad \tfrac{1}{2} (\beta_1 \beta_2^{-1} + \beta_2 \beta_1^{-1})\ H \otimes u(1,2);\end{aligned}$$

$(H \otimes u(1,3))^{\varphi_b}$ and $(H \otimes u(2,3))^{\varphi_b}$ are obtained from $(H \otimes u(1,2))^{\varphi_b}$ by the transformations above.

First we verify (7) in case a), where $\gamma_b = I$, $w(D + H \otimes c) = t(c)w$. This is readily seen to be satisfied: From the determinations of φ_b above, we see that each of $(0, 0, 0, T)^{\varphi_b}$, $(u, 0, 0, 0)^{\varphi_b}$ <u>etc</u>. annihilates W, as do the $(H \otimes u(i,j))^{\varphi_b}$, $1 \le i \le j \le 3$. Meanwhile, $(H \otimes \mathrm{diag}\ \{\gamma_1, \gamma_2, \gamma_3\})^{\varphi_b} = H \otimes \mathrm{diag}\ \{\gamma_1, \gamma_2, \gamma_3\}$, and these separate verifications give (7).

Now consider the case b), with $W = \mathrm{Der}(J) \times J_0$. With b as above, the effect of $\rho(b)$ on

$$(D,a) = ((u,v,y,T),\ \text{diag}\ \{\alpha_1,\alpha_2,\alpha_3\} + a_{12}(1,2) + a_{13}(1,3) + a_{23}(2,3))$$

is

$$\begin{aligned}&(t(b)D + 4D_{a,b},\ - bD + t(b)a)\\ &= ((t(b)u + 2(\beta_2 - \beta_1)a_{12},\ t(b)v + 2(\beta_3 - \beta_1)a_{13},\\ &\qquad t(b)y + 2\ (\beta_3 - \beta_2)a_{23},\ t(b)T),\\ &\qquad t(b)\ \text{diag}\ \{\alpha_1,\alpha_2,\alpha_3\} + (t(b)a_{12} + \tfrac{1}{2}(\beta_2 - \beta_1)u)(1,2)\\ &\quad + (t(b)a_{13} + \tfrac{1}{2}(\beta_3 - \beta_1)v)(1,3) + (t(b)a_{23} + \tfrac{1}{2}(\beta_3 - \beta_2)y(2,3)).\end{aligned}$$

From this one computes the effect on the same (D,a) of

$$\gamma_b = \rho(b)^3 - 3\rho(b)\ \rho(b^2) + 2\rho(b^3)$$

to be

$$(9)\qquad \begin{aligned}&(3(\beta_3(\beta_1^2 + \beta_2^2)u - 2\beta_3(\beta_1^2 - \beta_2^2)a_{12},\\ &\beta_2(\beta_1^2 + \beta_3^2)v - 2\beta_2(\beta_1^2 - \beta_3^2)a_{13},\\ &\beta_1(\beta_2^2 + \beta_3^2)y - 2\beta_1(\beta_2^2 - \beta_3^2)a_{23},\ 2\beta_1\beta_2\beta_3T),\\ &6\beta_1\beta_2\beta_3\ \text{diag}\ \{\alpha_1,\ \alpha_2,\ \alpha_3\}\\ &+ \tfrac{3}{2}\beta_3[(\beta_2^2 - \beta_1^2)u + 2(\beta_1^2 + \beta_2^2)a_{12}](1,2)\\ &+ \tfrac{3}{2}\beta_2[(\beta_3^2 - \beta_1^2)v + 2(\beta_1^2 + \beta_3^2)a_{13}](1,3)\\ &+ \tfrac{3}{2}\beta_1[(\beta_3^2 - \beta_2^2)y + 2(\beta_2^2 + \beta_3^2)a_{23}](2,3)).\end{aligned}$$

The relation (7) will be clear if we verify it separately for $x = (t,\ 0,\ 0,\ 0)$; $x = (0,\ 0,\ 0,\ S)$; $x = H \otimes \text{diag}\ \{\gamma_1,\ \gamma_2,\ \gamma_3\}$; $x = H \otimes t(1,2)$. When $x = (0,\ 0,\ 0,\ S)$, we have $x^{\varphi_b} = x$. The effect of x on our (D,a) is, from [24],

$$\begin{aligned}&(\tfrac{1}{2}(uS,\ vS^{\psi},\ yS^{\varphi},\ [TS]),\\ &\tfrac{1}{2}(a_{12}S)(1,2) + \tfrac{1}{2}(a_{13}S^{\psi})(1,3) + \tfrac{1}{2}(a_{23}S^{\varphi})(2,3)),\end{aligned}$$

and it is clear from (9) that its action commutes with that of γ_b.

Likewise, if $x = H \otimes \text{diag}\ \{\gamma_1,\ \gamma_2,\ \gamma_3\}$, $x^{\varphi_b} = x$, sending (D,a) to

$$\begin{aligned}&(2((\gamma_2 - \gamma_1)a_{12},\ (\gamma_3 - \gamma_1)a_{13},\ (\gamma_3 - \gamma_2)a_{23},\ 0),\\ &\tfrac{1}{2}(\gamma_2 - \gamma_1)u(1,2) + \tfrac{1}{2}(\gamma_3 - \gamma_1)v(1,3) + \tfrac{1}{2}(\gamma_3 - \gamma_2)y(2,3)).\end{aligned}$$

From this and (9) one sees easily that (7) holds.

For $x = (t, 0, 0, 0)$, we have $(D,a)\,x = ([D, (t, 0, 0, 0)], a(t, 0, 0, 0))$, which is seen from [24] to be

$$\Big(\tfrac{1}{2}(-tT, -ty, \bar{t}v, z \to 4(z,t)u - 4(z,u)t),$$
$$(a_{12},t)\operatorname{diag}\{-1, 1, 0\} + \tfrac{1}{2}(\alpha_1 - \alpha_2)t(1,2)$$
$$- \tfrac{1}{2}(ta_{23})(1,3) + \tfrac{1}{2}(\bar{t}a_{13})(2,3)\Big),$$

while

$$(D,a)(H \otimes t(1,2)) = (4D_{a,t(1,2)}, - t(1,2)D)$$
$$= (2((\alpha_1 - \alpha_2)t, - ta_{23}, -\overline{ta}_{13}, z \to 4(z, a_{12})t - 4(z,t)a_{12}),$$
$$- (t,u)\operatorname{diag}\{-1, 1, 0\} - \tfrac{1}{2}(tT)(1,2) - \tfrac{1}{2}(ty)(1,3) - \tfrac{1}{2}(\bar{t}v)(2,3)).$$

These formulas and those already given for $(t, 0, 0, 0)^{\varphi_b}$ and for $(H \otimes t(1,2))^{\varphi_b}$ make the remaining two verifications a straightforward matter. Thus we have verified the <u>second identities</u>:

<u>For</u> $\mathcal{W} = F$ or $\mathcal{W} = \mathrm{Der}(\mathcal{J}) \times \mathcal{J}_0$, <u>and for all</u> $a \in \mathcal{J}$, <u>all invertible</u> $b \in \mathcal{J}$, $\psi((E_{21} \otimes a)^4 (E_{12} \otimes b^2)^4)$ <u>annihilates</u> $\mathcal{W}$.

4. <u>FUNDAMENTAL REPRESENTATIONS FOR</u> $\mathfrak{sl}(2,\mathcal{J})$: <u>GENERAL IDENTITIES</u>.

The identities referred to are

$$w\psi((E_{21} \otimes a)^4 (E_{12} \otimes b)^4) = 0 \tag{10}$$

for the two L_0-modules a) and b), of weight $3\lambda_1$. When w and a are fixed, the mapping sending $b \in \mathcal{J}$ to the left-hand side of (10) is a polynomial mapping of $\mathcal{J}$ into the module $\mathcal{W}$. In the last section it was shown that this mapping is zero on squares of invertible elements of $\mathcal{J}$. Thus (10) will be established once we show that <u>squares of invertible elements are Zariski-dense</u> in $\mathcal{J}$. Here $\mathcal{J}$ is assumed a division algebra, so these are the same as the squares of non-zero elements of $\mathcal{J}$.

What we shall show is formally a little more, namely that <u>squares of elements</u> b <u>such that</u> R_b <u>is invertible are dense</u> in $\mathcal{J}$. For let $u_1, \ldots, u_n$ be a basis for $\mathcal{J}$, and let $p(X_1, \ldots, X_n)$ be a polynomial in n variables such that $p(\xi_1, \ldots, \xi_n) = 0$ whenever $\Sigma \xi_i u_i$ is the square of such an element b. If the multiplication in $\mathcal{J}$ is given by $u_i u_j = \Sigma_k \beta_{ijk} u_k$, then we have that

$$q(Y_1, \ldots, Y_n) = p(\sum_{i,j} Y_i Y_j \beta_{ij1}, \ldots, \sum_{i,j} Y_i Y_j \beta_{ijn})$$

vanishes on the set of elements b such that R_b is invertible. This set is evidently Zariski-open and contains 1, so is dense. Thus $q = 0$, and we wish to show $p = 0$. If not, we may assume p is of minimal degree among polynomials vanishing on all our squares, and let q be as above. Then each partial derivative $\frac{\partial q}{\partial Y_k}$ is also the zero-polynomial, that is,

$$0 = \frac{\partial q}{\partial Y_k} = \sum_\ell \frac{\partial p}{\partial X_\ell}\left(\sum_{i,j} Y_iY_j\beta_{ij1},\ldots,\sum_{i,j} Y_iY_j\beta_{ijn}\right)\left(2\sum_j Y_j\beta_{jk\ell}\right)$$

(from $u_iu_j = u_ju_i$, $\beta_{ij\ell} = \beta_{ji\ell}$ for all ℓ).

Specializing the right-hand side by substituting for the Y_i the coordinates of an element b such that R_b is invertible, say $b = \Sigma\xi_iu_i$, we have

$$0 = 2\sum_{\ell=1}^{n} \frac{\partial p}{\partial X_\ell}(b^2)\left(\sum_j \beta_{jk\ell}\,\xi_j\right)$$

for all k. Now the matrix of R_b relative to our basis is the matrix with (k, ℓ)-entry $\sum_j \beta_{jk\ell}\xi_j$, so the latter matrix is invertible. It follows that $\frac{\partial p}{\partial X_\ell}(b^2) = 0$ for all ℓ and for all b such that R_b is invertible. By the minimality of the degree of p, each $\frac{\partial p}{\partial X_\ell} = 0$, p is constant, and we have $p = 0$, a contradiction. Thus the general identity (10) is established.

It follows that the modules of a) and b) are $3\lambda_1$-admissible absolutely irreducible L_0-modules, so are the highest weight spaces of absolutely irreducible L-modules of highest weight $3\lambda_1$. These modules are clearly nonisomorphic, even upon splitting; by the fact that only two irreducible modules in the split case have highest weights restricting to $3\lambda_1$ on T, it follows from Proposition II.2 that our two modules exhaust the $3\lambda_1$-admissible irreducible L_0-modules. The constructions for $2\lambda_1$ and $3\lambda_1$ are now complete: <u>There is a unique irreducible L-module of highest weight $2\lambda_1$, namely the adjoint module of dimension 133, with highest weight space isomorphic to J as L_0-module ; there are two inequivalent irreducible L-modules of highest weight $3\lambda_1$, the highest weight spaces being the L_0-modules a) and b) of §2.</u> These L-modules are absolutely irreducible, and calculations in the split case give their respective dimensions as 56 and 912.

5. <u>REPRESENTATIONS OF $s\ell(2,J)$: COMPLETENESS.</u>

Let us decompose the tensor-product module $W \otimes W$, where $W = J$ as module for L_0 as in §2. Evidently $W \otimes W$ decomposes into the symmetric tensors $S^2(W)$, of dimension 378, and the skew tensors $\Lambda^2(W)$, of dimension 351. By passing to a splitting field, for instance, $[L_0L_0]$ is known to be the Lie algebra of all linear transformations of J skew with respect to the symmetric trilinear form (x, y, z) on J obtained by polarizing the cubic form

$$n(x) = t(x)^3 - 3t(x)t(x^2) + 2t(x^3).$$

From the trilinear form one obtains a nonzero map $\varphi: J \otimes J \to J^*$ sending $x \otimes y$ to the functional $z \to (x, y, z)$, with $\Lambda^2(W)$ in its kernel. The fact that $[L_0L_0]$ acts skewly with respect to (x, y, z) means that φ is a homomorphism of $[L_0L_0]$-modules, and induces a homomorphism $S^2(J) \to (J)^*$, which must be onto by the irreducibility of the latter. That is, if the contragredient $[L_0L_0]$-module J^* is made into an L_0-module by letting $H \otimes 1$ act as $4 \cdot \mathrm{Id.}$, J^* is a homomorphic image of the completely reducible L_0-module $W \otimes W$, hence, by §II.2, a $4\lambda_1$-admissible irreducible L_0-module. The absolute irreducibility of J implies that of J^*.

In a labeling consistent with that of Appendix 7, passage to a splitting field K for J makes $[L_0L_0]$ split of type E_6, with diagram

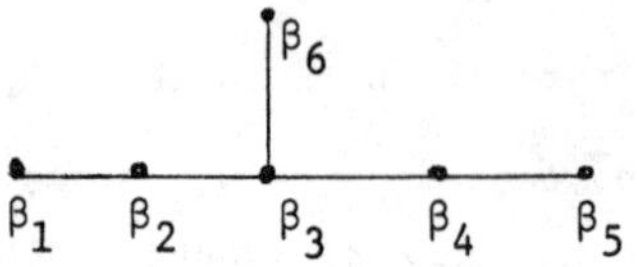

and J_K is the irreducible module M_{π_1} with highest weight π_1, J_K^* that with highest weight π_5. Now one determines routinely that

$$M_{\pi_1} \otimes M_{\pi_1} = M_{2\pi_1} \oplus M_{\pi_2} \oplus M_{\pi_5},$$

the three irreducible summands on the right satisfying $M_{\pi_2} = \Lambda^2(J_K)$ and $M_{2\pi_1} = \mathrm{Ker}(\varphi) \cap S^2(J_K)$, with φ as above. It follows that $\Lambda^2(W)$ and $\mathrm{Ker}(\varphi) \cap S^2(W)$, in the notation at the beginning of this section, are inequivalent absolutely irreducible $[L_0L_0]$-modules, not isomorphic to J^*, and thus that the L_0-module $W \otimes W$ decomposes into the three inequivalent $4\lambda_1$-admissible absolutely irreducible L_0-modules $\Lambda^2(W)$, $\mathrm{Ker}(\varphi) \cap S^2(W)$, J^*, of respective dimensions 351, 351, 27. These are, by Proposition II.2 and the remarks in §1, a complete set of $4\lambda_1$-admissible irreducible L_0-modules.

Next we show that if $W = J$ as above and if V is the $3\lambda_1$-admissible irreducible L_0-module $\mathrm{Der}(J) \times J_0$ of b) of §2, then all $5\lambda_1$-admissible irreducible L_0-modules occur in $V \otimes W$. As we have seen, V is the adjoint module for $[L_0L_0]$, so may be identified with a submodule of $\mathrm{End}_F(J)$.

In the notations above for the split case, V_K is the module M_{π_6} for

$[L_0L_0]_K$, and the tensor-product decomposition

$$V_K \otimes W_K = M_{\pi_6} \otimes M_{\pi_1} = M_{\pi_1+\pi_6} \oplus M_{\pi_4} \oplus M_{\pi_1}$$

holds. By Proposition II.2, if we can show that $V \otimes W$ contains the direct sum of three irreducible $[L_0L_0]$-modules, their sum must be $V \otimes W$, they must be absolutely irreducible, and (using a bound given in §1 on the number of highest weights restricting to $5\lambda_1$) they form a complete set of $5\lambda_1$-admissible irreducible L_0-modules.

For $x \in J$, $T \in [L_0L_0]$, let $\varphi(x,T)$ be the bilinear form on J defined by

$$\varphi(x,T)(y,z) = \frac{1}{2}(xT, y, z) + (x, yT, z).$$

It follows from the skewness of T with respect to the trilinear form that $\varphi(x,T)(y,z) + \varphi(x,T)(z,y) = 0$, so that $\varphi(x,T)$ identifies with an element of $\Lambda^2(J^*)$. Thus we obtain a linear mapping $\varphi: V \otimes W \mapsto \Lambda^2(J^*)$, which is directly verified to be a morphism of $[L_0L_0]$-modules. From the split case, $\Lambda^2(J^*)$ is an (absolutely) irreducible $[L_0L_0]$-module (of highest weight π_4, in our labeling of the diagram). Therefore either $\varphi = 0$ or φ maps $V \otimes W$ onto $\Lambda^2(J^*)$. To see that $\varphi \neq 0$, we may pass to the split case. From [23] it is easy to see that one can take $a, b, c \in O_K$, of (octonionic) traces zero, with $(ab, c) \neq 0$. Then taking $x = \text{diag}\,\{1, 0, 0\} \in J_K$, $T = R_{a(1,2)} \in [L_0L_0]_K$, $y = b(2,3)$, $z = c(1,3)$, one finds that $(x, yT, z) = 0$, while (xT, y, z) is a nonzero scalar multiple of $(ab, c) \neq 0$. Thus $\varphi(x,T)(y,z) \neq 0$ and $\varphi \neq 0$. Therefore, if $\Lambda^2(J^*)$ is made into an L_0-module by letting $H \otimes 1$ act by multiplication by 5, our φ is an L_0-module homomorphism of $V \otimes W$ onto $\Lambda^2(J^*)$, and $V \otimes W$ has an L_0-irreducible submodule isomorphic to $\Lambda^2(J^*)$.

There is a second submodule in $V \otimes W$, isomorphic to J as $[L_0L_0]$-module, obtained from the map sending $T \otimes x \mapsto xT$, an $[L_0L_0]$-homomorphism from $V \otimes W$ to J. Evidently J is not L_0-isomorphic to $\Lambda^2(J^*)$, nor is $J + \Lambda^2(J^*) = V \otimes W$. The existence of at least one more irreducible summand in $V \otimes W$ follows, and this is all we need to draw all the desired conclusions on $5\lambda_1$-admissible modules.

The study of $6\lambda_1$-admissible modules is more elaborate. First we consider those appearing in $V \otimes V$. Here V is the adjoint $[L_0L_0]$-module, absolutely irreducible of (split) highest weight π_6, and in the split case we know from [26], Appendix I that

$$M_{\pi_6} \otimes M_{\pi_6} = M_{2\pi_6} + M_{\pi_3} + M_{\pi_1+\pi_5} + M_{\pi_6} + M_0,$$

the summands having respective dimensions 2430, 2925, 650, 78, 1. As for $5\lambda_1$, we show that such a decomposition already occurs over F.

Here $S^2(V)$ and $\Lambda^2(V)$ are $[L_0L_0]$-submodules, of respective dimensions 3081 and 3003. The Killing form of $[L_0L_0]$ gives a homomorphism of $[L_0L_0]$-modules from $V \otimes V$ to F, vanishing on $\Lambda^2(V)$. Thus the <u>trivial $[L_0L_0]$-module</u> F <u>is a summand</u> of $S^2(V)$. Likewise the ordinary Lie bracket in $[L_0L_0]$ is a homomorphism of $V \otimes V$ to V, vanishing on $S^2(V)$, and the <u>adjoint module</u> $[L_0L_0]$ (= V) <u>is a summand of</u> $\Lambda^2(V)$.

Next we note that there is a linear mapping ψ of $V \otimes V$ into $\mathrm{End}_F(J)$ sending $S \otimes T$ to $ST + TS$, vanishing on $\Lambda^2(V)$. One sees at once that this is an $[L_0L_0]$-homomorphism, where the action of the subalgebra $[L_0L_0]$ on $\mathrm{End}_F(J)$ is the adjoint action. This module $\mathrm{End}_F(J)$ is equal to $FI \oplus [L_0L_0] \oplus M$, where M is a submodule of dimension 650. But from the identification of $\mathrm{End}_F(J)$ with $J^* \otimes J$ and from $J_K^* = M_{\pi_5}$, $J_K = M_{\pi_1}$, $M_{\pi_5} \otimes M_{\pi_1} = M_{\pi_1+\pi_5} \oplus M_{\pi_6} \oplus M_0$, we see as before that M must be absolutely irreducible. Thus, if the image of ψ meets M it contains M, and will be a submodule of $S^2(V)$.

Evidently all products ST of <u>two</u> elements of $[L_0L_0]$, as endomorphisms of J, are contained in $\psi(V \otimes V) + [L_0L_0]$, as are all linear combinations of such products. It is an easy matter to check (for instance, in the split case, where a verification is contained in §6 of [24]) that the linear span of such products has dimension greater than 79, so is not contained in $FI + [L_0L_0]$. It therefore meets M non-trivially, and it follows that the image of ψ contains M.

We now have $\Lambda^2(V) = [L_0L_0] \oplus R$, $S^2(V) = F \oplus M \oplus S$, where R and S are $[L_0L_0]$-submodules, of respective dimensions 2925 and 2430, and

$$V \otimes V = S \oplus R \oplus M \oplus [L_0L_0] \oplus F.$$

Comparing with the decomposition of $M_{\pi_6} \otimes M_{\pi_6}$, we see that S and R <u>are absolutely irreducible</u>.

Interpreted for L_0, the above says that <u>the tensor product</u> $V \otimes V$ <u>decomposes into five inequivalent</u> $6\lambda_1$<u>-admissible irreducible</u> L_0<u>-modules</u>, all of which are absolutely irreducible. These account for five of the seven irreducible L_K-modules whose highest weights restrict to $6\lambda_1$ on T. For the remaining ones we consider the third tensor power $W \otimes W \otimes W$ of the $2\lambda_1$-admissible module $W = J$.

Here $S^3(W)$ and $\Lambda^3(W)$ are direct summands of $\otimes^3 W$, of respective dimensions 3654 and 2925, and $\otimes^3 W \supset S^2(W) \otimes W \supset W^* \otimes W$, as in the case of $4\lambda_1$. Thus $\otimes^3 W$ contains the three absolutely irreducible summands M, $[L_0L_0]$,

F of dimensions 650, 78, 1, into which $W^* \otimes W$ decomposes. The invariant symmetric trilinear form (x, y, z) on $J = W$ shows that F occurs in $S^3(W)$.

Consider the mapping $\psi: \otimes^3 W \to \mathrm{End}(W)$ sending $x \otimes y \otimes z$ to the mapping $w \mapsto (x,y,w)z + (x,z,w)y + (y,z,w)x$. Then ψ is an $[L_0 L_0]$-morphism; we consider its restriction to $S^3(W)$. If we can show the image of $S^3(W)$ under ψ is not contained in $FI + [L_0 L_0]$, then this image must meet M and so must contain M. It will follow that $M \subset S^3(W)$.

If $U = \mu I + T$, $T \in [L_0 L_0] \subseteq \mathrm{End}_F(W)$, we have $(xU, y, z) + (x, yU, z) + (x, y, zU) = 3\mu(x, y, z)$ for all $x, y, z \in W$. Using the symmetry of ψ, it is enough to show that for some x and $y \in J$, $\psi(x \otimes x \otimes y)$ does not act as a similitude in this sense. It further suffices to assume $W = J$ is split. In the notations of [24], with e_1, e_2, e_3 the customary idempotents in J, let $x = e_1$, $y = e_2$. By §5 of [24] the typical element $a = \mathrm{diag}\,\{\alpha_1,\alpha_2,\alpha_3\} + a_{12}(1,2) + a_{13}(1,3) + a_{23}(2,3)$ of J is sent by $\psi(e_1 \otimes e_2 \otimes e_3)$ to a nonzero scalar K times $\alpha_3 e_1$, the value of K depending on a normalization of the form. Thus let U be the mapping sending a to $\alpha_3 e_1$, and let $u = e_3$, $v = b_{23}(2,3)$, $w = c_{23}(2,3)$ where b_{23} and c_{23} are in the split Cayley algebra O. Then (u,v,w) is seen from §5 of [24] to be zero, while
$(uU, v, w) + (u, vU, w) + (u, v, wU) = (e_1, v, w) = \nu(b_{23},c_{23})$, $\nu \neq 0$.
The last scalar product is that in O, and is not always zero. Therefore $M \subset S^3(W)$.

In the split case, if $w \in W$ is a highest weight vector belonging to the weight π_1 for $[L_0 L_0]$, then $w \otimes w \otimes w \in S^3(W)$ is a highest weight vector for $3\pi_1$, so that $S^3(W)$ contains the irreducible submodule of this highest weight, a module whose dimension is seen from Weyl's formula to be 3003. The dimension of $S^3(W)$ is $\binom{29}{3} = 3003 + 650 + 1$, so we have the full decomposition of $S^3(W)$ in the split case. In our setting, we have $S^3(W) = F \oplus M \oplus X$. It follows as before that X must be absolutely irreducible.

If $e_{\pm\beta_i}$ are root-vectors belonging to fundamental roots β_i and their negatives in $[L_0 L_0]_K$, with our labeling, and $w \in W_K$ is as above then

$$2w \otimes w \otimes we_{-\beta_1} - w \otimes we_{-\beta_1} \otimes w - we_{-\beta_1} \otimes w \otimes w$$

and

$$2w \otimes we_{-\beta_1} \otimes w - we_{-\beta_1} \otimes w \otimes w - w \otimes w \otimes we_{-\beta_1}$$

are linearly independent in $\otimes^3 W_K$, belong to the weight $3\pi_1 - \beta_1 = \pi_1 + \pi_2$, and are annihilated by all e_{β_i}. It follows that they are highest weight vectors for the $[L_0L_0]_K$-module of highest weight $\pi_1 + \pi_2$, whose dimension is 5824. This module therefore occurs in $\otimes^3 W_K$ with multiplicity at least two. In fact we can be more precise;

Consider the usual action of the symmetric group S_3 on $\otimes^3 W$ (or on $\otimes^3 W_K$). The operator $\frac{1}{6}S$, $S = \sum_{\sigma \in S_3} \sigma$, is central idempotent and maps $\otimes^3 W$ onto $S^3(W)$. The two elements last displayed above are in its kernel. Likewise the operator $\frac{1}{6}\Lambda$, $\Lambda = \sum_{\sigma \in S_3} \operatorname{sgn}(\sigma)\sigma$, is central, projects $\otimes^3 W$ onto $\Lambda^3(W)$ and has the two elements in its kernel. Now consider these elements of the group algebra of S_3,

$$\theta_1 = \frac{1}{3}(I + (1,2) - (1,3) - (1,2,3)),$$
$$\theta_2 = \frac{1}{3}(I + (1,3) - (1,2) - (1,3,2)).$$

Each is idempotent, and $\theta_1\theta_2 = 0 = \theta_2\theta_1 = \theta_i S = \theta_i\Lambda = \Lambda S$, $i = 1,2$. Moreover, $\theta_1 + \theta_2$ is central and $I = \frac{1}{6}S + (\theta_1 + \theta_2) + \frac{1}{6}\Lambda$. Clearly each of Λ, S, θ_1, θ_2 is an $[L_0L_0]$-endomorphism of $\otimes^3 W$ and $\otimes^3 W$ is the direct sum of their four images. Of the two elements of $\otimes^3 W_K$ displayed in the last paragraph, the former is in the image of θ_1 and the latter in the image of θ_2. Thus each of these images contains the irreducible $[L_0L_0]_K$-module of highest weight $\pi_1 + \pi_2$ and dimension 5824.

The maps $\theta_1(1,2)\theta_2$ and $\theta_2(1,3)\theta_1$ send, respectively, the image of θ_1 onto that of θ_2 and <u>vice versa</u>, and their composites are θ_1 and θ_2. Thus they afford inverse isomorphisms of $[L_0L_0]$-modules between the image of θ_1 and that of θ_2. The operator Λ sends $w \otimes we_{-\beta_1} \otimes we_{-\beta_1}e_{-\beta_2} \neq 0$ in $\otimes^3 W_K$ to a non-zero element of $\Lambda^3(W_K)$, belonging to the weight $3\pi_1 - 2\beta_1 - \beta_2 = \pi_3$ and annihilated by all e_{β_i}. Thus $\Lambda^3 W_K$ contains the irreducible $[L_0L_0]_K$-module of highest weight π_3, whose dimension is 2925. From the fact that $2925 = \binom{27}{3} = \dim \Lambda^3(W)$, we see that <u>$\Lambda^3(W)$ is an absolutely irreducible $[L_0L_0]$-module</u>.

We have seen that the adjoint module $[L_0L_0]$ occurs in $\otimes^3 W$, but not in either of $S^3(W)$ or $\Lambda^3(W)$. <u>Thus it occurs in both the image of θ_1 and that of θ_2</u>. We next show that <u>the 650-dimensional module M occurs</u> in both of these. For this, note that the mapping η of $\otimes^3 W$ to $\operatorname{End}(W)$ sending a typical generator $x \otimes y \otimes z$ to $w \to (x, y, w)z$ is a morphism of $[L_0L_0]$-modules and sends

$$x \otimes y \otimes z + y \otimes x \otimes z - z \otimes y \otimes x - z \otimes x \otimes y,$$

in the image of θ_1, to the mapping

$$w \to 2(x, y, w)z - (z, y, w)x - (z, x, w)y.$$

Similarly, the typical generator

$$x \otimes y \otimes z + z \otimes y \otimes x - y \otimes x \otimes z - y \otimes z \otimes x$$

of the image of θ_2 goes under η' to the mapping

$$x \to 2(x,z,w)y - (y,z,w)x - (y, x, w)z,$$

where η' sends $x \otimes y \otimes z$ to $w \to (x, z, w)y$.

As before, we see upon passing to the split case that when $x = y = e_1$ and $z = e_2$ neither mapping of W_K into itself is in $KI + [L_0L_0]_K$. That the images (now over F) contain M follows as before, so that for $i = 1, 2$ the image of θ_i contains both M and $[L_0L_0]$. Each of these images has dimension $\frac{1}{2}(27^3 - \binom{27}{3} - \binom{29}{3}) = 6552$; if E is either image, we thus have $E = [L_0L_0] \oplus M \oplus G$, while $E_K = [L_0L_0]_K \oplus M_K \oplus G_K$ contains an irreducible submodule of highest weight $\pi_1 + \pi_2$ and dimension 5824 = dim G. It follows that G is absolutely irreducible.

Now we have that $\otimes^3 W$ contains six nonisomorphic irreducible modules X, G, M, $[L_0L_0]$, $\Lambda^3(W)$, F, of respective dimensions: 3003; 5824; 650; 78; 2925; 1. All these are absolutely irreducible. Together with the irreducible module S of dimension 2430 contained in $S^2(V)$, they make up a set of seven inequivalent $6\lambda_1$-admissible L_0-modules, all absolutely irreducible. Again it follows by the bound of seven in §1 and by Proposition II.2 that these seven are the complete set of $6\lambda_1$-admissible irreducible L_0-modules.

We have now proved

THEOREM X.1. For J an exceptional central Jordan division algebra over F and $L = s\ell(2,J)$, a fundamental set of weights and associated L_0-modules, in the sense of §V.7, consists of the weights $2\lambda_1$ and $3\lambda_1$, and:

for $2\lambda_1$, of the single $2\lambda_1$-admissible L_0-module J of §2; the associated L-module is the adjoint module;

for $3\lambda_1$, of the two $3\lambda_1$-admissible L_0-modules F and $\mathrm{Der}(J) \times J_0$ of a) and b) of §2; the associated L-modules are absolutely irreducible of dimension 56 and 912.

6. REMARKS ON $s\ell(2,C)$, C A CUBIC EXTENSION.

Here C is an extension field of F, $[C:F] = 3$. The F-Lie algebra $L = s\ell(2,C)$ has C as centroid; upon splitting C, L becomes a direct sum of three ideals, each isomorphic to $s\ell(2,K)$, where K is the splitting field. Over F, $T = FH$, $H = \mathrm{diag}\{-1, 1\}$, is a maximal split torus in L, $L_0 = CH$. A fundamental set of dominant integral functions in the split case consists of the three fundamental ones, one for each simple summand, and all these have restriction λ_1 to T.

From Proposition I.3, we know that any irreducible L-module of highest weight λ_1 is actually a module for the C-Lie algebra $s\ell(2,C)$, so has the structure of C-vector space C^2, with the usual action of $s\ell(2,C)$. The centralizer of the L-action on C^2 is a field, namely C itself, so this module C^2 is irreducible as module for L over F. Its highest weight-space $C(\cong u_2 \otimes C)$ is the unique λ_1-admissible L_0-module, the action of $H \otimes a$ on $u_2 \otimes b$ sending the latter to $u_2 \otimes ab$. We shall call on this fact in §XI.3.

CHAPTER XI

EXCEPTIONAL TYPES IV: RELATIVE TYPE G_2.

In [26] it is shown that all central simple Lie algebras over F of relative type G_2 arise from "Tits' second construction" involving the split octonion algebra (for which we use the notation $\mathcal{O}$ again -- there is no risk of confusion with the division algebras of Chapter IX) and a Jordan division algebra $\mathcal{A}$ such that every element $a \in \mathcal{A}$ is a root of a generic monic cubic polynomial $p_a(X)$, the coefficients in $p_a(X)$ being forms of the appropriate degree in the coordinates of a.

From the classification of simple Jordan algebras in [17], one sees that $\mathcal{A}$ must be one of the algebras i) - v) listed at the start of Chapter X, except that the center of $\mathcal{A}$ may now be an extension field of F. When the condition that every element be a root of a cubic polynomial is imposed, we see that the centroid must either be F or a cubic extension field C of F. In the latter case, only $\mathcal{A} = C$ is possible, since otherwise a primitive element of $C[a]$, where $a \in \mathcal{A}$, $a \notin C$, cannot be cubic over F. Thus our possibilities for $\mathcal{A}$ are narrowed to $\mathcal{A} = C$ and to the central division algebras of Chapter X. The effect of the cubic restriction on the latter is as follows:

i) Here the central associative division algebra $\mathcal{A}$ must satisfy a cubic. There are only two possibilities: $\mathcal{A} = F$ and $[\mathcal{A}: F] = 9$. In the former case the Lie algebra is split of type G_2, and its representation theory will be assumed known (see, e.g., [15], Chapter VII).

ii) If $\mathcal{B}$ is a central associative division algebra with involution of first kind, then $[\mathcal{B}: F] = 2^{2n}$ for some n; if $b \in \mathcal{B}$, then $[F[b]:F] = 2^m$ for some $m \leq n$, and b cannot be cubic over F. This case does not occur unless $\mathcal{B} = F$, when we have the split case of i).

iii) The forms which are the coefficients in $p_a(X)$ may be evaluated at elements of $\mathcal{A}_K$, where K is any extension field, and one sees by polarization that $p_a(a) = 0$ for all $a \in \mathcal{A}_K$. When K is taken to split the center of the involutorial algebra of second kind, this algebra becomes the sum of a central simple associative algebra and its opposite, and the fixed elements under the involution are isomorphic to the central simple algebra. Apart from the trivial case where this algebra is a field (and the Lie algebra is split of type G_2), it must be of degree 3. Thus the central associative division algebra with involution of second kind is of dimension 9 over its center Z.

iv) In the Jordan algebra A of a quadratic form over F, we have $[F[a]: F] = 2$ for all $a \in A$, $a \notin F$; this case does not occur.

v) The exceptional Jordan division algebra J, of dimension 27, has been seen in Chapter X to satisfy our conditions.

1. THE ALGEBRAS $G_2(A)$ AND THEIR FUNDAMENTAL WEIGHTS.

As with J, there is a linear trace function $t(a)$ on A with $t(1) = 3$ and for all $a \in A$,

$$p_a(X) = X^3 - t(a)X^2 + \frac{1}{2}(t(a)^2 - t(a^2))X - \frac{1}{6}(t(a)^3 - 3t(a)t(a^2) + 2t(a^3)).$$

The kernel A_0 of t contains the image of A under each derivation of A and is stable under the bilinear product $a \circ b = ab - \frac{1}{3}t(ab)1$, where juxtaposition denotes the commutative Jordan product. All derivations of A are inner, that is, are linear combinations of the $D_{a,b}: c \to (ca)b - (cb)a$, with $D_{a,1} = 0$ for all a.

Likewise in $\mathcal{O}$, we have a "generic trace" t, which is linear from $\mathcal{O}$ to F, with $t(1) = 2$ and kernel $\mathcal{O}_0$. Each $u \in \mathcal{O}$ is a root of $p_u(X) = X^2 - t(u)X + \frac{1}{2}(t(u)^2 - t(u^2))$. We define $(u,v) = \frac{1}{2}t(uv)$, a nondegenerate symmetric bilinear form of maximal Witt index on $\mathcal{O}$ and on $\mathcal{O}_0$. The inner derivation $D_{u,v}$ for $u, v \in \mathcal{O}$ is defined as in §VIII.1, and all derivations are linear combinations of these. We have: $D_{v,u} = -D_{u,v}$: $D_{1,u} = 0$; all derivations map $\mathcal{O}$ into $\mathcal{O}_0$; $(uD, v) + (u, vD) = 0$, for all derivations D of $\mathcal{O}$; $[uv] = uv - vu \in \mathcal{O}_0$, for all $u, v \in \mathcal{O}$.

The Lie algebra $L = G_2(A)$ has the underlying vector space

$$L = \mathrm{Der}(A) \oplus \mathrm{Der}(\mathcal{O}) \oplus (A_0 \otimes \mathcal{O}_0),$$

with bracket as follows:

$\mathrm{Der}(A)$ and $\mathrm{Der}(\mathcal{O})$ are subalgebras, centralizing each other. For $u \in \mathcal{O}_0$, $a \in A_0$, $E \in \mathrm{Der}(A)$, $D \in \mathrm{Der}(\mathcal{O})$, we have $[u \otimes a, D] = uD \otimes a$; $[u \otimes a, E] = u \otimes aE$. For $u, v \in \mathcal{O}_0$; $a, b \in A_0$,

$$[u \otimes a, v \otimes b] = (u, v)D_{a,b} + \frac{1}{12} t(ab)D_{u,v} + \frac{1}{2}[uv] \otimes (a \circ b).$$

In $\mathcal{O}$ there are non-zero idempotents $\varepsilon_1, \varepsilon_2$, $\varepsilon_1\varepsilon_2 = 0$, $\varepsilon_1 + \varepsilon_2 = 1$. The spaces $\varepsilon_i \mathcal{O} \varepsilon_j$, $i \neq j$, are (well-defined and) 3-dimensional, totally isotropic with respect to the form (u,v), and orthogonal to $\varepsilon_1, \varepsilon_2$. We may take a basis v_1, v_2, v_3 for $\varepsilon_1 \mathcal{O} \varepsilon_2$ and a basis v_4, v_5, v_6 for $\varepsilon_2 \mathcal{O} \varepsilon_1$

such that the only non-zero scalar products among these are $(v_i, v_{7-i}) = 1$, $1 \le i \le 6$. A basis for $\mathrm{Der}(\mathcal{O})$ then consists of; <u>eight</u> transformations mapping ε_1 and ε_2 to 0, and on the space $\varepsilon_1 \mathcal{O} \varepsilon_2 + \varepsilon_2 \mathcal{O} \varepsilon_1$ having matrices $E_{ij} - E_{7-j,7-i}$, $1 \le i \ne j \le 3$, and $D_1 = E_{22} - E_{11} - E_{55} + E_{66}$, $D_2 = E_{33} - E_{22} - E_{44} + E_{55}$ relative to the basis $v_1, \ldots, v_6$; and of the <u>six</u> inner derivations D_{v_i,ε_1} $(= - D_{v_i,\varepsilon_2})$, $1 \le i \le 6$.

The subspace $FD_1 + FD_2$ of $\mathrm{Der}(\mathcal{O})$ is a maximal split torus T in L, with centralizer $L_0 = \mathrm{Der}(A) \otimes T \otimes ((\varepsilon_2 - \varepsilon_1) \otimes A_0)$. Relative to T, L breaks up into six root-spaces L_α of dimension one, each having as basis an $E_{ij} - E_{7-j,7-i}$, $i \ne j$, as above; and six root-spaces isomorphic as linear space to A, each having the form $FD_{v_i,\varepsilon_1} + v_i \otimes A_0$. We set $L_\alpha = F(E_{13} - E_{46})$, $L_{\alpha_2} = FD_{v_4,\varepsilon_1} + v_4 \otimes A_0$, $H_1 = D_1 + D_2$, $H_2 = -D_1 - 2D_2$; then α_1 and α_2 form a fundamental system of roots of type G_2, with $\alpha_i(H_i) = 2$, $\alpha_1(H_2) = -3 = 3\alpha_2(H_1)$.

The information on the splitting of L contained in Appendices 8 and 9 shows that if (λ_1, λ_2) is a basis for T^* dual to (H_1, H_2), then each irreducible L-module results by Cartan multiplication, as in Chapter II, from those with highest weights:

i) – iii) λ_1, $j\lambda_2$, $1 \le j \le 3$;

v) λ_1, $j\lambda_2$, $2 \le j \le 6$;

vi) λ_1, λ_2 .

Here the cases i), iii), v) correspond to the admissible choices of A as labeled before, and vi) is the case where A is a cubic extension field C of F. If K is a splitting field for L, the numbers of highest weights (in a consistent ordering) of irreducible representations for L_K with the above as restrictions to T are seen from Appendices 8 and 9 to be:

i) – iii) λ_1: <u>one</u>; λ_2; <u>two</u>; $2\lambda_2$: <u>five</u>; $3\lambda_2$: <u>nine</u>.

v) λ_1: <u>one</u>; $2\lambda_2$: <u>one</u>; $3\lambda_2$: <u>two</u>; $4\lambda_2$: <u>three</u>; $5\lambda_2$: <u>three</u>; $6\lambda_2$: <u>seven</u>.

vi) λ_1: <u>one</u>; λ_2: <u>three</u>.

2. AN EMBEDDING OF $s\ell(2,A)$ IN $G_2(A)$.

We fix a realization of $G_2(A)$ more precisely by specifying some data in $\mathcal{O}$, following §6 of [23], using a version of Zorn's vector-matrix realization. A basis f_1, f_2, f_3 for ordinary 3-space F^3 is chosen, with the usual scalar and vector products $f\cdot g$ resp. $f \wedge g$ as for $\mathbb{R}^3$. Then $\mathcal{O}$ is realized as 2 by 2 matrices of the form

$$\begin{pmatrix} \alpha & f \\ g & \beta \end{pmatrix}$$

with $\alpha, \beta \in F; f,g \in F^3$. The trace is $\alpha + \beta$, and the product is

$$\begin{pmatrix} \alpha & f \\ g & \beta \end{pmatrix}\begin{pmatrix} \alpha' & f' \\ g' & \beta' \end{pmatrix} = \begin{pmatrix} \alpha\alpha' + f\cdot g' & \alpha f' + \beta' f - g\wedge g' \\ \alpha' g + \beta g' + f\wedge f' & \beta\beta' + g\cdot f' \end{pmatrix}$$

The usual unit matrix 1 serves as unit element, and $\varepsilon_1 = \begin{pmatrix} 0 & 0 \\ 0 & 1 \end{pmatrix}$ and $1 - \varepsilon_1 = \varepsilon_2$ are a pair of orthogonal idempotents. With

$$v_i = \begin{pmatrix} 0 & 0 \\ 2f_{4-i} & 0 \end{pmatrix}, \quad 1 \le i \le 3, \quad \text{and} \quad v_{7-i} = \begin{pmatrix} 0 & f_{4-i} \\ 0 & 0 \end{pmatrix}, \quad 1 \le i \le 3,$$

we have basis elements for $\mathcal{O}$ as in §1.

In the notations of [23] for derivations of $\mathcal{O}$, the derivation $(f_1,0,0)$ spans the same one-dimensional space as does D_{v_4,ε_1} of §1, and the following hold:

$$[(f_1, 0, 0), H_2] = 2(f_1, 0, 0);$$

(1)
$$\begin{pmatrix} 0 & f_1 \\ 0 & 0 \end{pmatrix} H_2 = 2\begin{pmatrix} 0 & f_1 \\ 0 & 0 \end{pmatrix};$$

$$\begin{pmatrix} -1 & 0 \\ 0 & 1 \end{pmatrix}(f_1, 0, 0) = -\begin{pmatrix} 0 & f_1 \\ 0 & 0 \end{pmatrix};$$

$$\left[\begin{pmatrix} 0 & f_1 \\ 0 & 0 \end{pmatrix}, \begin{pmatrix} -1 & 0 \\ 0 & 1 \end{pmatrix}\right] = 2\begin{pmatrix} 0 & f_1 \\ 0 & 0 \end{pmatrix}$$

$$D_{\begin{pmatrix} 0 & f_1 \\ 0 & 0 \end{pmatrix}, \begin{pmatrix} -1 & 0 \\ 0 & 1 \end{pmatrix}} = 4(f_1, 0, 0)$$

Using $s\ell(2,A)$ to denote in general the algebra obtained, as in Chapter X, from A and the Koecher-Tits construction, we define a linear mapping φ_0 from $s\ell(2,A)_0$ to $G_2(A)_0$ as follows:

The elements of $s\ell(2,A)_0$ have the form $D + (H \otimes a)$, $D \in \mathrm{Der}(A)$, $a \in A$, $H = \mathrm{diag}\{-1, 1\}$. The image of $D + (H \otimes a)$ under φ_0 is to be

$$D + \frac{1}{3}\, t(a)H_2 + 2 \begin{pmatrix} -1 & 0 \\ 0 & 1 \end{pmatrix} \otimes a_0,$$

where $a_0 = a - \frac{1}{3}\, t(a)\, 1 \in A_0$. From

$$\left(\begin{pmatrix} -1 & 0 \\ 0 & 1 \end{pmatrix}, \begin{pmatrix} -1 & 0 \\ 0 & 1 \end{pmatrix}\right) = 1$$

and the formulas of §1 it is clear that φ_0 is a homomorphism, indeed a monomorphism, of Lie algebras. The image is easily seen to be the subspace $[L_{\alpha_2}, L_{-\alpha_2}]$ of $G_2(A)$, in the notation of §1. The objective here is to extend φ_0 to an isomorphism of $s\ell(2,A)$ onto the subalgebra $L_{\alpha_2} + L_{-\alpha_2} + [L_{\alpha_2}\ L_{-\alpha_2}] = M$ of $G_2(A)$.

Explicitly, we send $E_{12} \otimes a \in s\ell(2,A)$ to $\frac{1}{3}\, t(a)(f_1, 0, 0) + \begin{pmatrix} 0 & f_1 \\ 0 & 0 \end{pmatrix} \otimes a_0$, obtaining a linear isomorphism φ_1 of $E_{12} \otimes A$, the root-space "L_α" in $s\ell(2,A)$, onto L_{α_2}. The formulas (1) suffice to show that for $x \in E_{12} \otimes A$, $y \in s\ell(2,A)_0$, $[xy]^{\varphi_1} = [x^{\varphi_1}, y^{\varphi_0}]$, so that φ_1 <u>is an isomorphism of</u> $s\ell(2,A)_0$-modules. Again we extend φ_0 and φ_1 to an isomorphism of $s\ell(2,A)$ onto M by appeal to the version of Allison's extension theorem invoked in §VIII.4. For this we must show only that M is simple with FH_2 as maximal split torus, having $[L_{\alpha_2}\ L_{-\alpha_2}]$ as its centralizer, and that φ_0 satisfies the appropriate compatibility condition for representatives of the Weyl groups.

If N is a non-zero ideal of M, the stability of N under ad H_2 shows that N is the sum of its intersections with the three spaces L_{α_2}, $L_{-\alpha_2}$, $[L_{\alpha_2}\ L_{-\alpha_2}]$. Now $L_0 = G_2(A)_0 = [L_{\alpha_2}\ L_{-\alpha_2}] + FH_0$, where $H_0 \in T$, $\alpha_2(H_0) = 0$, and L_0 acts irreducibly in each L_β; it follows that if $N \cap L_{\alpha_2}$ (or $N \cap L_{-\alpha_2}$) is non-zero then L_{α_2} (or $L_{-\alpha_2}$) $\subseteq N$. But then $[L_{\alpha_2}\ L_{-\alpha_2}] \subseteq N$, $H_2 \in N$ and $N = M$. Otherwise we have $N \subseteq [L_{\alpha_2}\ L_{-\alpha_2}]$, $[NL_{\alpha_2}] = 0 = [NL_{-\alpha_2}]$, and N is central in M. Thus N consists of elements $H_2 \otimes a$, centralizing $L_{\pm\alpha_2}$, and one verifies from the definitions of §1 that this is only possible if $a = 0$. Therefore M is simple. That FH_2 is a maximal split torus in M with centralizer $[L_{\alpha_2} L_{-\alpha_2}]$ is evident from the properties of $G_2(A)$ in §III.5 of [26].

For the condition of compatibility involving the Weyl groups, it will be enough to show that the canonical automorphisms $\omega = \exp(\mathrm{ad}\, E_{12})\exp(\mathrm{ad}(-E_{21}))\exp(\mathrm{ad}\, E_{12})$ of $s\ell(2,A)$, interchanging $E_{12} \otimes A$ and $E_{21} \otimes A$, and $\omega' = \exp(\mathrm{ad}(f_1, 0, 0))\exp(\mathrm{ad}\, 4(0, f_1, 0))\exp(\mathrm{ad}(f_1, 0, 0))$ of M, interchanging L_{α_2} and $L_{-\alpha_2}$, are such that on $s\ell(2,A)_0$, $\eta = \varphi_0\, \omega'\, \varphi_0^{-1}\, \omega^{-1}$ is the identity.

Since all of $\varphi_0, \omega', \omega$ fix $D \in \mathrm{Der}(A)$, it is clear that η does. On $H \otimes a$, we have

$$(H \otimes a)^{\varphi_0} = \frac{1}{3} t(a)H_2 + 2(\varepsilon_1 - \varepsilon_2) \otimes a_0$$

$$H_2^{\omega'} = -H_2, \quad ((\varepsilon_1 - \varepsilon_2) \otimes a_0)^{\omega'} = -(\varepsilon_1 - \varepsilon_2) \otimes a_0,$$

so

$$(H \otimes a)^{\varphi_0 \omega' \varphi_0^{-1}} = -H \otimes a = (H \otimes a)^{\omega},$$

so that $(H \otimes a)^{\eta} = H \otimes a$. It now follows that φ_0 and φ_1 admit an (unique) extension to an isomorphism φ of $s\ell(2,A)$ onto M, necessarily mapping $E_{21} \otimes A$ onto $L_{-\alpha_2}$.

The embedding φ of $s\ell(2,A)$ in G_2 A) has $L_0 = \varphi(s\ell(2,A)_0) + FH_1$, where $H_1 \in T$ is central in L_0, and therefore satisfies all the conditions of §VIII.4, used there as well as in §§IX.7-9, to show that (in present notation) the irreducible λ-admissible L_0-modules are the irreducible $\lambda \circ \varphi$ -admissible $s\ell(2,A)_0$-modules.

3. CONSTRUCTION OF FUNDAMENTAL REPRESENTATIONS.

From $(H \otimes 1)^{\varphi} = H_2 \otimes 1$, we have $\lambda_1 \circ \varphi = 0$, $\lambda_2 \circ \varphi = $ "λ_1", the quotation-marks referring to conventions for $s\ell(2,A)$. The consequences of previous considerations are now listed separately for the cases i), iii), v), vi):

i) Here $A = \mathcal{D}$, a cubic central associative division algebra, and it has been shown in §V.2a of [26] that $s\ell(2,A)$ is the algebra $s\ell(2,\mathcal{D})$ of Chapter V, with $s\ell(2,A)_0 = [\mathcal{D}\mathcal{D}] + H \otimes \mathcal{D}$ as there, $E_{ij} \otimes A = E_{ij} \otimes \mathcal{D}$ the root-spaces for $(i,j) = (1,2)$ or $(2,1)$. The only 0-admissible irreducible $s\ell(2,A)_0$-module is the trivial module F, the only "λ_1"-admissible ones are two whose underlying space is $\mathcal{D}$, the action of $s\ell(2,A)_0$ being given in

§V.3, and all "$k\lambda_1$"-admissible irreducible L_0-modules for $k > 1$ are obtained from these by Cartan multiplication (Theroem V.2). There follows:

THEOREM XI.1. *Let A be the Jordan algebra of a central associative division algebra $\mathcal{D}$ of degree 3 over F. A fundamental set for $G_2(A)$ consists of the weights λ_1, λ_2 and of irreducible λ-admissible L_0-modules $\mathcal{W}$ as follows:*

$\lambda = \lambda_1$: $\mathcal{W}$ *of dimension one, with*

$$w(D + \mu_1 H_1 + \mu_2 H_2 + (\varepsilon_2 - \varepsilon_1) \otimes a_0) = \mu_1 w.$$

$\lambda = \lambda_2$: *Two modules $\mathcal{W}_1$, $\mathcal{W}_2$ each with underlying space $\mathcal{D}$, and with*

$$w_1 g = (\mu_2 - d' - a_0)w_1,$$
$$w_2 g = w_2(\mu_2 + d' - a_0),$$

where $g = D + \mu_1 H_1 + \mu_2 H_2 + (\varepsilon_2 - \varepsilon_1) \otimes a_0$ *as above, with* $d' \in [\mathcal{D}\mathcal{D}]$ *such that* $D = ad(d')$.

The corresponding fundamental modules for $L = G_2(A)$ are: for $\lambda = \lambda_1$, the adjoint module, which is absolutely irreducible of dimension 78. For $\lambda = \lambda_2$, two modules, each of which splits into three isomorphic modules, each of highest weight π_1 in one case, of highest weight π_5 in the other, upon passage to a splitting field. Thus each irreducible L-module of highest weight π_2 has dimension $3 \cdot 27 = 81$. The two modules are contragredient to one another.

iii) From §V.8.a of [26], the algebra $\mathfrak{sl}(2,A)$ is that investigated in Chapter VI, with the restrictions that $n = 1$ and that $[\mathcal{D}: Z] = 9$. As an element of the corresponding "L_0" in the algebra of Chapter VI, our general element g of $\mathfrak{sl}(2,A)_0$ above is identified by the isomorphism of [26], V.7.a, with a matrix $g = - a_1^* E_{11} + a_1 E_{22}$, with $\tau^-(a_1) = 0$. The element H identifies with that g for which $a_1 = 1$. The unique "λ_1"-admissible irreducible "L_0"-module is $\mathcal{D}$ itself, the action of g in matrix form on $\mathcal{D}$ being right multiplication by a_1. If we denote by g^η the image in $L_0 \subseteq G_2(A)$ under the map φ of the element of $\mathfrak{sl}(2,A)_0$ corresponding to $g = - a_1^* E_{11} + a_1 E_{22}$, every element of L_0 can be uniquely written as $g^\eta + \mu_1 H_1$ for such a "g" and for $\mu_1 \in F$. The consequences for $G_2(A)$ now read:

THEOREM XI.2. *Let the Jordan division algebra A be the set of fixed elements under an involution of second kind in a division algebra $\mathcal{D}$ of* degree 3 over its center Z, a quadratic extension of F. Then a fundamental set for $G_2(A)$ consists of the weights λ_1 and λ_2 and, for each of these, of a single irreducible admissible L_0-module $\mathcal{W}$. For λ_1, $\mathcal{W} = F$, the action of $g^{\eta} + \mu_1 H_1$ being multiplication by μ_1. For λ_2, $\mathcal{W} = \mathcal{D}$, the action of $g^{\eta} + \mu_1 H_1$ being right multiplication by a_1.

The corresponding irreducible L-module for λ_1 is the adjoint module, of dimension 78; for λ_2, it becomes upon splitting the sum of three copies of the module of highest weight π_1 and three copies of that of highest weight π_5. Thus its dimension is $6\cdot 27 = 162$.

v) Here the necessary information has been derived in Chapter X, in a notation consistent with that of this chapter. The translation is immediate from Theorem X.1:

THEOREM XI.3. Let J be an exceptional central simple Jordan division algebra over F. Then a fundamental set for $G_2(J)$ consists of the weights λ_1, $2\lambda_2$, $3\lambda_2$, and, with $g = D + \mu_1 H_1 + \mu_2 H_2 + (\varepsilon_2 - \varepsilon_1) \otimes a_0$ the general element of L_0, of admissible irreducible L_0-modules as follows:

For λ_1: $\mathcal{W} = F$, $wg = \mu_1 w$.

For $2\lambda_2$: $\mathcal{W} = J$, $wg = wD + w(2\mu_2 - a_0)$.

For $3\lambda_2$: There are two modules:

$\mathcal{W} = F$, $wg = 3\mu_2 w$; and

$\mathcal{W} = \mathrm{Der}(J) \times J_0$, with

$$(E, b_0)g = ([E,D] + 3\mu_2 E - 2D_{b_0,a_0},\; b_0 D + \frac{1}{2} a_0 E + 3\mu_2 b_0) .$$

The irreducible L_0-modules in question are absolutely irreducible, so the same holds for the irreducible L-modules of these highest weights. Upon splitting the latter become those of respective highest weights π_1, π_7, π_2, π_8 (in our labeling for E_8), and have respective dimensions: 248 (the adjoint module); 3875; 30,380; 147,250.

vi) Here $\mathrm{Der}(C) = 0$, and the elements of L_0 have the form
$$g = \mu_1 H_1 + \mu_2 H_2 + (\varepsilon_2 - \varepsilon_1) \otimes a_0 = \mu_1 H_1 + (H \otimes (\mu_2 - \frac{1}{2} a_0))^{\varphi_b} .$$
A λ_1-admissible irreducible L_0-module must be annihilated by the image of L_0, so must be one-dimensional. This observation and those of §X.6 combine to give

THEOREM XI.4. Let $L = G_2(C)$, C a cubic extension field of F. A fundamental set for L consists of the two weights λ_1, λ_2 and, for each of these, of a unique admissible irreducible L_0-module $\mathcal{W}$, as follows:

For λ_1: $\mathcal{W} = F$, $wg = \mu_1 g$.

For λ_2: $\mathcal{W} = C$, $wg = \mu_2 w - \frac{1}{2} a_0 w$.

The corresponding L-module of highest weight λ_1 is absolutely irreducible and is the adjoint module, of dimension 28. Upon extension of the base field to split L, the module of highest weight λ_2 decomposes into three nonisomorphic irreducible modules, each of dimension 8. Thus the irreducible L-module of highest weight λ_2 has dimension 24.

APPENDICES: SPLITTING INFORMATION.

1. $\mathfrak{sl}(n,\mathcal{D})$.

Let K be an extension field of F such that $\mathcal{D}_K \cong M_d(K)$, and let e_{ij}, $1 \le i, j \le d$ be a set of d by d matrix units in $\mathcal{D}_K$. Then $\mathfrak{sl}(n,\mathcal{D})_K \cong \mathfrak{sl}(nd, K)$, a split algebra of type A_{nd-1}, viewed as the subspace of $M_n(M_d(K))$ consisting of those n by n matrices

$$(a_{ij}) = \sum_{i,j=1}^{n} a_{ij}E_{ij}, \text{ where the } a_{ij} \text{ are in } M_d(K), \text{such that } \sum_{i=1}^{n} tr(a_{ii})=0.$$

Here the E_{ij} are n by n matrix units. The elements $H_i = E_{i+1,i+1}-E_{ii}$ that form a basis for T identify here with $I_dE_{i+1,i+1} - I_dE_{ii}$, where I_d is the d by d identity matrix, and a Cartan subalgebra H of $\mathfrak{sl}(n,\mathcal{D})_K$ consists of all the nd by nd diagonal matrices of trace zero, with basis $(e_{j+1,j+1}- e_{jj})E_{ii}$, $1 \le i \le n$, $1 \le j < d$, and $e_{11}E_{i+1,i +1} - e_{dd}E_{ii}$, $1 \le i < n$.

A fundamental system of roots $\gamma_1,\ldots, \gamma_{nd-1}$ relative to H has as root-vectors,

for γ_{kd+j}, $0 \le k < n$, $0 < j < d$: $e_{j,j+1}E_{k+1,k+1}$;

for γ_{id}, $1 \le i < n$: $e_{d1}E_{i,i+1}$. The diagram is

A_{nd-1}: $\underset{\gamma_1}{\bullet}$——$\underset{\gamma_2}{\bullet}$ .... •——$\underset{\gamma_{nd-1}}{\bullet}$.

Of these two sets, the former centralizes T, while $\gamma_{id}|T = \alpha_i$, in the notations of §V.1.

From [3], Planche I, p. 250, fundamental weights relative to H associated with the fundamental system $\gamma_1,\ldots,\gamma_\ell$ are given by

$$\pi_j = \frac{1}{nd} \left(\sum_{k=1}^{j} k(nd-j)\gamma_k + \sum_{k=j+1}^{nd-1} j(nd-k)\gamma_k \right),$$

so that if $j = id$,

$$\pi_{id}|T = \frac{1}{nd} \left(\sum_{\ell=1}^{i} \ell d(nd-id)\alpha_\ell + \sum_{\ell=i+1}^{n-1} id(nd-\ell d)\alpha_\ell \right) = d\lambda_i.$$

If $j = id + (d-k)$, $0 < k < d$, $0 \le i < n$,

$$\pi_j\big|_T = \frac{1}{nd}\left(\sum_{\ell=1}^{i} \ell d(nd-id-d+k)\alpha_\ell + \sum_{\ell=i+1}^{n-1} (id+(d-k))(nd-\ell d)\alpha_\ell\right)$$

$$= \frac{k}{n}\left(\sum_{\ell=1}^{i} \ell(n-i-1)\alpha_\ell + \sum_{\ell=i+1}^{n-1} i(n-\ell)\alpha_\ell\right)$$

$$+ \frac{d-k}{n}\left(\sum_{\ell=1}^{i+1} \ell(n-i-1)\alpha_\ell + \sum_{\ell=i+2}^{n-1} (i+1)(n-\ell)\alpha_\ell\right) = k\lambda_i+(d-k)\lambda_{i+1}.$$

In particular, for $n = 2$, the $\pi_j|_T$ for $j = 1, 2, \ldots, 2d-1$ are $\lambda_1, 2\lambda_1, \ldots, d\lambda_1, (d-1)\lambda_1, \ldots, \lambda_1$, respectively. Those non-negative integral combinations of $\pi_1,\ldots,\pi_{2d-1}$ whose restrictions to T are equal to $k\lambda_1$, $1 \le k < d$, are obtained by taking a pair of non-negative integers r, s, $r + s = k$, a partition $P \leftrightarrow (u_1,\ldots,u_r)$ of r and a partition $P' = (v_1,\ldots,v_s)$ of s, and forming $\pi = \Sigma u_i\pi_i + v_i\pi_{2d-i}$. These are distinct. When $k = d$, the same applies, except that for $r = 0$, $s = d$ and for $r = d$, $s = 0$, the respective partitions $P = (0,\ldots,0, 1)$ and $P' = (0,\ldots,0, 1)$ give π_d.

In Chapter V, we have also used information about the splitting of certain modules. One set of these, in §V.3, occurred for $\mathfrak{sl}(2,\mathcal{D})$; we had $r < d$, so that $S^r(\mathcal{D}^{opp})$ gave rise to $p(r)$ inequivalent $r\lambda_1$-admissible L_0-modules. These are minimal right ideals in $S^r(\mathcal{D}^{opp})$, so are right $S^r(\mathcal{D}^{opp})$-submodules of $\otimes^r\mathcal{D}^{opp}$. For general $n \ge 2$, the unique λ_1-admissible irreducible L_0-module $\mathcal{D}^{opp}$ is the highest weight-space of the irreducible $\mathfrak{sl}(n,\mathcal{D})$-module consisting of column n-vectors from $\mathcal{D}$, the action of a matrix (a_{ij}) from $\mathfrak{sl}(n,\mathcal{D})$ on a vector (b_i) being the <u>negative</u> of <u>left</u> multiplication. The vectors annihilated by all positive root-vectors (<u>i.e.</u> by all strictly upper triangular matrices) are those such that $b_i = 0$ for all $i > 1$. If (b_i) is such a column vector, the first entry in $-(\Sigma a_jE_{jj})(b_i)$ is $-a_1b_1 = -b_1\rho_1(a_1)$, where ρ_1 is our mapping of $\mathcal{D}(=\mathcal{D}^{opp})$ into $S^1(\mathcal{D}^{opp})$.

Upon passing to the splitting field K, minimal left ideals in $\mathcal{D}_K$ whose direct sum is $\mathcal{D}_K$ are obtained by fixing j, $1 \le j \le d$, and taking all $d \times d$ K-matrices which are zero outside the j-th column, <u>i.e.</u>, $\sum_{i=1}^{d} Ke_{ij}$, for each such j. The $\mathfrak{sl}(n,\mathcal{D})$-module of highest weight λ_1 decomposes into a direct sum of d isomorphic $\mathfrak{sl}(n,\mathcal{D})_K$-modules, consisting of all column-vectors with entries in the j-th left ideal above, for $j = 1,\ldots,d$. In the split setting, a highest weight vector for the j-th of these is the column vector with e_{1j} in the first position and zeros elsewhere. Under the negative of left multiplication by a diagonal matrix in

$M_n(M_d(K))$, say $\sum_{i=1}^{n} \sum_{j=1}^{d} x_{ij}e_{jj}E_{ii}$, this is sent to $-x_{11}$ times itself. From our definition of the γ_i, this means that the highest weight as $s\ell(n,\mathcal{D})_K$-module is π_1. That is, the module R for $s\ell(n,\mathcal{D})$ with highest weight λ_1 and highest weight-space $\mathcal{D}^{opp}$ splits over K into a sum of irreducible modules of highest weight π_1. Denote by M one of these. Then $(\otimes^r R)_K$ has the same irreducible constituents as does $\otimes_K^r M$, and all weights of this module are of the form $r\pi_1 - \sum_{i=1}^{nd-1} m_i\gamma_i$, m_i non-negative integers.

For $j > r$, the coefficient of γ_j in the expression from [3] for $r\pi_1$ is $\frac{1}{nd} r(nd-j)$, while the coefficient of γ_j in π_j is $\frac{1}{nd} j(nd-j) > \frac{1}{nd} r(nd-j)$. It follows that π_j cannot be a weight of $\otimes_K^r M$, and in fact that all weights of $\otimes_K^r M$ must be non-negative integral combinations of $\pi_1, \ldots, \pi_r$. Because $\otimes^r \mathcal{D}^{opp}$ is an L_0-submodule of $\otimes^r R$, all T-weights in this module must be restrictions of such combinations. This is the property used in §V.3.

2. ALGEBRAS OF RELATIVE TYPES C (FIRST KIND).

Here the Lie algebra L, its maximal split torus T and its roots are as in Chapter V. The involutorial division algebra $(\mathcal{D}, *)$ has degree d a power of 2. Upon passage to a splitting field K, it is isomorphic (as involutorial algebra) to one of the following - here d = 2s:

Orthogonal type: All d by d matrices, the involution being reflection in the nonprincipal diagonal.

Symplectic type: All d by d k-matrices, the involution sending

$$\left(\begin{array}{c|c} A & B \\ \hline C & D \end{array}\right) \text{ to } \left(\begin{array}{c|c} D' & -B' \\ \hline -C' & A' \end{array}\right),$$

where M' is the reflection of M in its nonprincipal diagonal.

With e_{ij} the usual d × d matrix units in $\mathcal{D}_K$, E_{ij} those in $M_{2n}(\mathcal{D}) \subseteq M_{2n}(\mathcal{D}_K)$, a Cartan subalgebra H containing T consists of the linear span of all $-e_{jj}^*E_{ii} + e_{jj}E_{2n+1-i,2n+1-i}$, $1 \le j \le d$, $1 \le i \le n$, thus of all $-e_{kk}E_{ii} + e_{d+1-k,d+1-k}E_{2n+1-i,2n+1-i}$, $1 \le k \le d$, $1 \le i \le n$. When the involution is of orthogonal type, a fundamental system of roots $\gamma_1, \ldots, \gamma_{nd}$ relative to H has as root-vectors,

for γ_{kd+j}, $0 \le k < n$, $0 < j < d$:

$$-e_{j,j+1}E_{k+1,k+1} + e_{d-j,d-j+1}E_{2n-k,2n-k};$$

for γ_{id}, $1 \leq i < n$: $- e_{d,1}E_{i,i+1} + e_{d,1}E_{2n-i,2n-i+1}$;

for γ_{nd}: $e_{d,1}E_{n,n+1}$. In this order, these form a system of type C_{nd}:

Evidently T is centralized by root-vectors above where the index is not divisible by d; for the others, one has $\gamma_{id}|_T = \alpha_i$, in the notation of §V.1. From Planche III, p. 254 of [3], the corresponding fundamental weights π_i to $\{\gamma_1, \ldots, \gamma_{nd}\}$ are given by

$$\pi_i = \sum_{j=1}^{i-1} j\gamma_j + i(\gamma_i + \ldots + \gamma_{nd-1} + \frac{1}{2}\gamma_{nd}),$$

and one sees from this that $\pi_{di}|T = d\lambda_i$, $1 \leq i \leq n$, while $\pi_{id+(d-k)}|T$, $0 \leq i < n$, $0 < k < d$, is equal to $k\lambda_i + (d-k)\lambda_{i+1}$, with $\lambda_0 = 0$ by definition.

When the involution is of symplectic type, the root-vectors for γ_{kd+j}, $0 \leq k < n$, $0 \leq j < d$, are as above; for γ_{id}. $1 \leq i < n$, one uses

$$e_{d1}E_{i,i+1} + e_{d1}E_{2n-i,2n-i+1};$$

for γ_{nd}, one takes $(- e_{d-1,1} + e_{d2})E_{n,n+1}$ (here $d \geq 2$). The resulting system is of type D_{nd}, with diagram

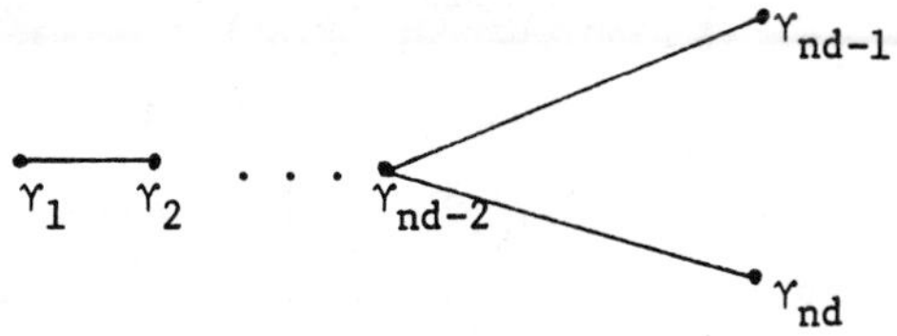

Again $\gamma_{id}|_T = \alpha_i$ while $\gamma_j|_T = 0$ if $d \nmid j$.

From Planche IV, p. 256 of [3], the corresponding fundamental weights π_i have

$$\pi_i = \sum_{j=1}^{i-1} j\gamma_j + i(\gamma_i + \ldots + \gamma_{nd-2} + \frac{1}{2}\gamma_{nd-1} + \frac{1}{2}\gamma_{nd}), \quad i \leq nd-2;$$

$$\pi_{nd-1} = \frac{1}{2}\left(\sum_{j=1}^{nd-2} j\gamma_j\right) + \frac{nd}{4}\gamma_{nd-1} + \frac{nd-2}{4}\gamma_{nd};$$

$$\pi_{nd} = \frac{1}{2}\left(\sum_{j=1}^{nd-2} j\gamma_j\right) + \frac{nd-2}{4}\gamma_{nd-1} + \frac{nd}{4}\gamma_{nd}.$$

Again one verifies easily that $\pi_{id}|_T = d\lambda_i$, $1 \leq i < n$, while

$\pi_{nd|T} = \frac{d}{2}\gamma_n$, and that, for $0 \le i < n$, $0 < k < d$, $id + (d-k) < nd-1$, $\pi_{id+(d-k)|T} = k\lambda_i + (d-k)\lambda_{i+1}$ $(\lambda_0 = 0)$. Finally

$$\pi_{nd-1|T} = \frac{1}{2}\left(\sum_{i=1}^{n-1} di\,\alpha_i\right) + \frac{nd-2}{4}\alpha_n$$

$$= \left(\sum_{i=1}^{n-1} i\alpha_i + \frac{n-1}{2}\alpha_n\right) + \left(\frac{d}{2} - 1\right)\left(\sum_{i=1}^{n-1} i\alpha_i + \frac{n}{2}\alpha_n\right)$$

$$= \lambda_{n-1} + \left(\frac{d}{2} - 1\right)\lambda_n.$$

Under some circumstances, one wants the Cartan subalgebra H to be defined over F. This can be assumed to be the case here and in §1: All that is necessary in §1 is to take H to be the diagonal-matrices in $\mathfrak{sl}(n,\mathcal{D})$ with entries in a maximal subfield E of $\mathcal{D}$, then to take K a field such that $E \otimes K$ has $[E:F] = d$ orthogonal primitive idempotents; in §2, one must further choose E to be stable under the involution, a choice which is always possible.

In §V.5, we have considered an $(s-1)\lambda_1$-admissible module $\mathcal{D}_1$ when $n = 1$, $d = 2s$, and the involution in $\mathcal{D}$ is of symplectic type. The underlying space of $\mathcal{D}_1$ is $\mathcal{D}$, the action of the general element $g = -a^*E_{11} + aE_{22}$ of L_0 on $b \in \mathcal{D}$ sending b to $s\tau(a)b - ba^*$. The fact used in Chapter V was a consequence of the fact that in the $(L_0)_K$-module $(\mathcal{D}_1)_K$ all irreducible submodules have highest weight π_{d-1}.

As right $\mathcal{D}_K$-module, $\mathcal{D}_K$ is the sum of d minimal right ideals, corresponding to the matrix-rows $\sum_{j=1}^{d} Ke_{ij}$, $1 \le i \le d$. Evidently each of these is an irreducible $(L_0)_K$-submodule of $(\mathcal{D}_1)_K$, and all are isomorphic. The elements $h_1,\ldots,h_d$ forming a basis for H with $\pi_i(h_j) = \delta_{ij}$ are $h_i = (e_{j+1,j+1} - e_{jj})E_{11} + (e_{d-j+1,d-j+1} - e_{d-j,d-j})E_{22}$, $1 \le j < d$, and $h_d = (e_{22} + e_{11})E_{22} - (e_{d-1,d-1} + e_{dd})E_{11}$. The subset of $\sum_{j=1}^{d} Ke_{ij}$ annihilated by all positive root-vectors in L_0, i.e., by all $e^*_{k\ell}$, $k < \ell$, is the same as that annihilated upon right multiplication by all such $e_{k\ell}$, so consists of Ke_{id}. Now in the action of $(L_0)_K$ on $(\mathcal{D}_1)_K$,

$e_{id}h_j = -e_{id}(e_{d-j+1,d-j+1} - e_{d-j,d-j})^* = e_{id}(e_{j+1,j+1} - e_{jj}) = \delta_{j+1,d}e_{id}$ if $j < d$, while $e_{id}h_d = s\tau(e_{11} + e_{22})e_{id} - e_{id}(e_{11} + e_{22})^* =$ $s\cdot\frac{2}{d}\,e_{id} - e_{id}(e_{dd} + e_{d-1,d-1}) = e_{id} - e_{id} = 0$. Thus e_{id} belongs to the weight π_{d-1}, and our claim is established.

3. ALGEBRAS OF RELATIVE TYPE C (SECOND KIND).

Upon splitting, the division algebra $\mathcal{D}$ with involution $a \to a^*$ of second kind and center Z, $[\mathcal{D}:Z] = d^2$, is isomorphic as involutorial algebra to the set of all $2d$ by $2d$ matrices of the form

$$\left(\begin{array}{c|c} A & 0 \\ \hline 0 & B \end{array}\right)$$

where A and B are d by d matrices, the involution being reflection in the non-principal diagonal (of length $2d$). The Lie algebra L_K, with L of Chapter VI, identifies with the set of all $2n$ by $2n$ $\mathcal{D}_K$-matrices, where K is our splitting field, satisfying the conditions given in §VI.1. Here the a_{ij} should be regarded as elements of $\mathcal{D}_K$ as above, thus as 2d-rowed matrices, and the involution $a \to a^*$ is that above: Over K, the mapping $\tau:\mathcal{D}_K \to Z_K$ sends

$$\left(\begin{array}{c|c} A & 0 \\ \hline 0 & B \end{array}\right) \quad \text{to} \quad \left(\begin{array}{c|c} \frac{1}{d}\,\mathrm{tr}(A)I_d & 0 \\ \hline 0 & \frac{1}{d}\,\mathrm{tr}(B)I_d \end{array}\right)$$

$$= \frac{1}{2d}\,\mathrm{tr}(A+B)I_{2d} + \frac{1}{2d}\,\mathrm{tr}(A-B)\left(\begin{array}{c|c} I_d & 0 \\ \hline 0 & -I_d \end{array}\right),$$

and, over K, our $\zeta \in Z$, $\zeta^* = -\zeta$, becomes a scalar multiple of $\left(\begin{array}{c|c} I_d & 0 \\ \hline 0 & -I_d \end{array}\right)$. Thus the condition $\sum_{i=1}^{n} \bar{\tau}(a_{ii}) = 0$ of §VI.1 becomes, over K, $\sum_{i=1}^{n} \mathrm{tr}(a_{ii}^{(1)} - a_{ii}^{(2)}) = 0$, where $a = \left(\begin{array}{c|c} a^{(1)} & 0 \\ \hline 0 & a^{(2)} \end{array}\right)$, for $a \in \mathcal{D}_K$, the $a^{(j)}$ being d by d matrices.

A Cartan subalgebra H of L_K containing our T of §VI.1 may be obtained by taking all diagonal matrices (now $4nd$ by $4nd$) in L_K, and is defined over F if we start with the set of all $2n \times 2n$ diagonal matrices in L with entries in a $*$-stable maximal subfield of $\mathcal{D}$, which subfield we then split by K. We use the notation e_{ij} for $2d$ by $2d$ matrix units. The subalgebra H has dimension $2nd-1$; we next list a fundamental system of roots for L_K with respect to H.

For $0 \le i < n$, $1 \le j < d$,

$$- e^*_{j,j+1}E_{i+1,i+1} + e_{j,j+1}E_{2n-i,2n-i}$$
$$= - e_{2d-j,2d+1-j}E_{i+1,i+1} + e_{j,j+1}\,E_{2n-i,2n-i}$$

is a root-vector; denote the associated root by γ_{id+j}. For $1 \le i < n$,

$- e^*_{d,1}E_{i,i+1} + e_{d,1}E_{2n-i,2n+1-i} =$

$- e_{2d,d+1}E_{i,i+1} + e_{d,1}E_{2n-i,2n+1-i}$ is a root-vector, whose associated root we denote by γ_{id}. The resulting set $\gamma_1,\ldots,\gamma_{nd-1}$ of roots is linearly independent, has the property that no $\gamma_i - \gamma_j$ is a root if $i \neq j$, and $\gamma_i + \gamma_j$ is a root if and only if $|i-j| = 1$.

Again for $0 \leq i < n$, $1 \leq j < d$,

$- e_{j,j+1}E_{i+1,i+1} + e^*_{j,j+1} E_{2n-i,2n-i}$

$= - e_{j,j+1}E_{i+1,i+1} + e_{2d-j,2d-j+1} E_{2n-i,2n-i}$ is a root-vector, whose associated root we designate by $\gamma_{(2n-i)d-j}$: likewise, for $1 \leq i < n$, $- e_{d,1}E_{i,i+1} + e_{2d,d+1}E_{2n-i,2n+1-i}$ is a root-vector, whose root is denoted by $\gamma_{(2n-i)d}$. The results of this paragraph are $nd-1$ roots $\gamma_{nd+1}, \gamma_{nd+2},\ldots, \gamma_{2nd-1}$, satisfying the same conditions as the set $\{\gamma_1,\ldots, \gamma_{nd-1}\}$ of the preceding paragraph. Moreover, these sets are (disjoint and) linearly independent when taken together, and no sum or difference of a root from the first set and one from the second set is a root.

The linkage between the two sets is supplied by γ_{nd}, the root to which the following belongs: $(e_{d1} + e^*_{d1})E_{n,n+1} = (e_{d,1} + e_{2d,d+1})E_{n,n+1}$. Then $\gamma_1,\ldots, \gamma_{2nd-1}$ form a fundamental system of roots for L_K relative to H, of type A_{2nd-1}, with diagram

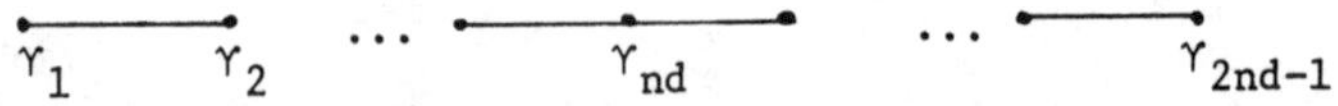

From the fact that T is spanned by the $E_{ii} - E_{2n+1-i,2n+1-i}$, it is clear that each π_j with $d \nmid j$ vanishes on T. On the other hand, from the choice of the roots $\alpha_1,\ldots, \alpha_n$ in §VI.1 one sees that $\gamma_{id}|T = \alpha_i = \gamma_{(2n-i)d}|T$, $1 \leq i \leq n$. This is the required information for use in Chapter VI.

4. <u>ALGEBRAS OF TYPE</u> B.

The notations here are those of §VII.1. If $q = 2r$ is <u>even</u>, we take an extension K of F such that the form (u,v) is of maximal index r on B_K. If $q = 2r-1$ is <u>odd</u>, we take K so that (u,v) is of maximal index on $Ke_{n+1} + B_K$. In the latter case, a basis for $Ke_{n+1} + B_K$ may be taken to consist of $e_{n+1},\ldots, e_{n+2r}$, with $(e_{n+j+1}, e_{n+2r-j}) = 1$, $1 \leq j \leq r-1$; with $(e_{n+1}, e_{n+1}) = 1 = -(e_{n+2r}, e_{n+2r})$; with all other pairs being orthogonal. We form a new basis $f_1,\ldots,f_{2(n+r)}$ for V_K by setting $f_j = e_j$ for $j \neq n+1$,

$n + 2r$; $f_{n+1} = e_{n+1} + e_{n+2r}$; $f_{n+2r} = \frac{1}{2}(e_{n+1} - e_{n+2r})$.

Still with $q = 2r-1$, a Cartan subalgebra H of L_K containing T consists of the skew transformations having diagonal matrices relative to $f_1, \ldots, f_{2(n+r)}$, i.e., of K-linear combinations of the matrices $e_{2(n+r)+1-i,2(n+r)+1-i} - e_{ii}$, $1 \le i \le n + r$. For $1 \le i \le n$, these are the K-linear extensions to V_K of the basis $E_{2n+q+2-i,2n+q+2-i} - E_{ii}$ for T. The Cartan subalgebra H may be taken to be defined over F by forming it by field extension from the set of elements of L stabilizing each of r orthogonal two-dimensional subspaces of $Fe_{n+1} + B$, with sum $Fe_{n+1} + B$.

A fundamental system of roots for L_K relative to H consists of roots γ_i, $1 \le i < n + r$, to which $e_{i,i+1} - e_{2(n+r)-i,2(n+r)+1-i}$ belongs, together with γ_{n+r}, to which $e_{n+r-1,n+r+1} - e_{n+r,n+r+2}$ belongs. The diagram of this system is

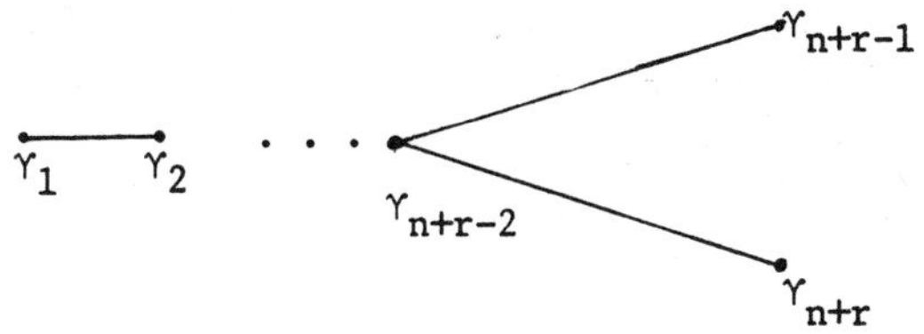

of type D_{n+r}.

From the basis for T consisting of the H_i, $1 \le i \le n$, as in §VII.1, we see that $\gamma_j|_T = 0$ for $j \le n$, while $\gamma_i|_T = \alpha_i$, $1 \le i \le n$.

When $q = 2r$, we take a basis $e_{n+2}, \ldots, e_{n+2r+1}$ for B_K, with $(e_i, e_j) = \delta_{i+j,2(n+r)+3}$, and $H \subseteq L_K$ is to be the diagonal matrices relative to the basis $e_1, \ldots, e_{2(n+r)+1}$ for V_K. (As above, H may be taken to be defined over F.) A basis for H consists of the basis $E_{2n+2r+2-i,2n+2r+2-i} - E_{ii}$, $1 \le i \le n$, for T, together with the

$$e_{n+2r+2-j,n+2r+2-j} - e_{n+j+1,n+j+1}, \quad 1 \le j \le r.$$

A fundamental system of roots $\gamma_1, \ldots, \gamma_{n+r}$ relative to H has: for $1 \le i < n$, $e_{i,i+1} - e_{2n+2r+1-i,2n+2r+2-i}$ belonging to γ_i;

$$e_{n,n+2} - e_{n+2r+1,n+2r+2} \quad \text{belonging to} \quad \gamma_n;$$

for $1 \le j < r$, $e_{n+j+1,n+j+2} - e_{n+2r+1-j,n+2r+2-j}$ belonging to γ_{n+j};

$$e_{n+r+1,n+1} - e_{n+1,n+r+2} \quad \text{belonging to} \quad \gamma_{n+r}.$$

Again it is clear that $\gamma_j|_T = 0$ for $j > n$; meanwhile $\gamma_i|_T = \alpha_i$, for $i \le n$. The diagram of $\gamma_1, \ldots, \gamma_{n+r}$ is of type B_{n+r}:

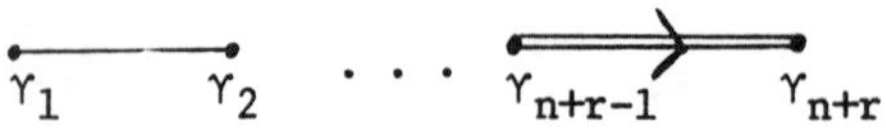

.

This completes the splitting information of §VII.1. The fundamental weights π_i given there are to be found in Planches II and IV, pp. 252 and 256, of [3].

5. ALGEBRAS OF RELATIVE TYPE F_4.

The settings and notations are those of Chapters VIII and IX. Upon performing a finite extension of base field, say to K, we may assume $\mathcal{D}_K = K \oplus K$, $M_2(K)$, or the split octonions over K, as in Chapter XI, §2; we use the notations to be found there, in this last case. Thus ε_1 and ε_2 are primitive idempotents in $\mathcal{O}_K$ identified with the respective Zorn vector-matrices diag {0, 1} , diag {1, 0} . In $M_2(K)$ we denote the same ordinary 2 by 2 matrices as ε_1 and ε_2, and we may use the same labels if $K \oplus K$ is viewed as the 2 by 2 diagonal K-matrices. When $\mathcal{D} = \mathcal{O}$, we take v_1, v_2, v_3 to be the basis of §XI.2 for $\varepsilon_1 \mathcal{D}_K \varepsilon_2$, and v_6, v_5, v_4 the dual basis there for $\varepsilon_2 \mathcal{D}_K \varepsilon_1$. When $\mathcal{D} = \mathcal{Q}$, we take a dual basis with respect to the trace form of $\mathcal{Q}_K$ consisting of $v_1 \in \varepsilon_1 \mathcal{Q}_K \varepsilon_2$ and $v_2 \in \varepsilon_2 \mathcal{Q}_K \varepsilon_1$.

A splitting Cartan subalgebra $\mathcal{H}_0$ of $\mathrm{Der}(\mathcal{D}_K)$ has (K-)-diagonalizable action in $\mathcal{D}_K$ and annihilates the subspace $K\varepsilon_1 + K\varepsilon_2 = \mathcal{E}$. The orthogonal space $\mathcal{E}^\perp$ of $\mathcal{E}$ with respect to the trace form is $\varepsilon_1 \mathcal{D}_K \varepsilon_2 + \varepsilon_2 \mathcal{D}_K \varepsilon_1$, which is zero if $\mathcal{D}_K = K \oplus K$ and has basis $\{v_1, v_2\}$ resp. $\{v_1, \ldots, v_6\}$ in the other cases. When $\mathcal{D}_K = K \oplus K$, $\mathcal{H}_0 = 0$; when $\mathcal{D}_K = \mathcal{Q}_K$, $\mathcal{H}_0$ stabilizes $\mathcal{E}^\perp$ and has a basis consisting of a derivation D with matrix diag {-1, 1} in $\mathcal{E}^\perp$ relative to the basis v_1, v_2. When $\mathcal{D}_K = \mathcal{O}_K$, a basis for $\mathcal{H}_0$ consists of D_1 and D_2, with matrices in $\mathcal{E}^\perp$ relative to $v_1, \ldots, v_6$ as follows:

$$D_1: \ \mathrm{diag}\,\{-1, 1, 0, 0, -1, 1\} ;$$

$$D_2: \ \mathrm{diag}\,\{ 0, -1, 1, -1, 1, 0\} .$$

For consistency, we use $U_1 = \mathrm{diag}\,\{-1, 1, 0\}$ and $U_2 = \mathrm{diag}\,\{0, -1, 1\}$ in the exceptional split Jordan algebra $\mathcal{J}$ of Chapter VIII, in fact in $\mathcal{J}_0$. Let $\mathcal{H}$ be the subspace of $\mathcal{L}_K$ spanned by:

$H_1, H_2, H_3, H_4, U_1 \otimes (\varepsilon_2 - \varepsilon_1), U_2 \otimes (\varepsilon_2 - \varepsilon_1)$, if $\mathcal{D}_K = K \oplus K$;

by the above six and D if $\mathcal{D}_K = \mathcal{Q}_K$;
by the above six and D_1, D_2 if $\mathcal{D}_K = \mathcal{O}_K$.

Then $\mathcal{H}$ is a commutative subalgbera of $\mathcal{L}_K$. We have the following root-vectors relative to H in L_K:

i) All derivations of J of the form $(0, 0, 0, E_{ij} - E_{9-j,9-i})$, $i \neq j$, $1 \leq i, j \leq 4$; of the form $(0, 0, 0, E_{i,9-j} - E_{j,9-i})$, and all of the form $(0, 0, 0, E_{9-j,i} - E_{9-i,j})$, $1 \leq i \leq j \leq 4$; there are 24 of these.

ii) All elements $D_{u_i(1,2),U_1} \pm 2u_i(1,2) \otimes (\varepsilon_2 - \varepsilon_1)$, $1 \leq i \leq 8$; all $D_{u_i(1,3),U_1} \pm u_i(1,3) \otimes (\varepsilon_2 - \varepsilon_1)$, $1 \leq i \leq 8$; all

$D_{u_i(2,3),U_1} \pm u_i(2,3) \otimes (\varepsilon_2 - \varepsilon_1)$, $1 \leq i \leq 8$. There are 48 of these.

When $\mathcal{D}_K = K \oplus K$, there are no more root-vectors; the seventy-two above constitute a complete set relative to the Cartan subalgebra H.

iii) All $u_i(1,2) \otimes v_j$, $u_i(1,3) \otimes v_j$, $u_i(2,3) \otimes v_j$, $1 \leq i \leq 8$, with $1 \leq j \leq 2$ if $\mathcal{D}_K = \mathcal{Q}_K$, $1 \leq j \leq 6$ if $\mathcal{D}_K = \mathcal{O}_K$. There are 48 of these if $\mathcal{D} = \mathcal{Q}$, 144 of them if $\mathcal{D} = \mathcal{O}$.

iv) All $U_1 \otimes v_i + 2U_2 \otimes v_i - D_{v_i,\varepsilon_1}$; all $U_1 \otimes v_i - U_2 \otimes v_i - D_{v_i,\varepsilon_1}$; all $2U_1 \otimes v_i + U_2 \otimes v_i + D_{v_i,\varepsilon_1}$, for $i = 1$ if $\mathcal{D} = \mathcal{Q}$ and for $1 \leq i \leq 3$ if $\mathcal{D} = \mathcal{O}$. There are 3 of these if $\mathcal{D} = \mathcal{Q}$, 9 if $\mathcal{D} = \mathcal{O}$.

v) All $2U_1 \otimes v_i + U_2 \otimes v_i - D_{v_i,\varepsilon_1}$; $U_1 \otimes v_i - U_2 \otimes v_i + D_{v_i,\varepsilon_1}$; $U_1 \otimes v_i + 2U_2 \otimes v_i + D_{v_i,\varepsilon_1}$, for $i = 2$ if $\mathcal{D} = \mathcal{Q}$; $4 \leq i \leq 6$ if $\mathcal{D} = \mathcal{O}$. Again there are 3 resp. 9 of these.

If $\mathcal{D} = \mathcal{Q}$, the above are 126 root-vectors and are a complete set relative to H. If $\mathcal{D} = \mathcal{O}$, there are six more, namely

vi) Six derivations of $\mathcal{O}_K$ annihilating ε_1 and ε_2, their matrices relative to the basis $v_1,\ldots,v_6$ for $E^{\perp}$ being

$$E_{ij} - E_{7-j,7-i}, \quad i \neq j, \quad 1 \leq i, j \leq 3.$$

Altogether when $\mathcal{D} = \mathcal{O}$ the above constitute a set of 240 root-vectors in L_K relative to H. Fundamental systems of roots, with corresponding root-vectors, and with the sequence consisting of their values at the following bases for H are given below:

When $\mathcal{D}_K = K \oplus K$; basis for H: $U_1 \otimes (\varepsilon_2 - \varepsilon_1)$, $U_2 \otimes (\varepsilon_2 - \varepsilon_1)$, H_1, H_2, H_3, H_4;

When $\mathcal{D} = \mathcal{Q}$; basis for H : D_1, then the six elements above.

When $\mathcal{D} = \mathcal{O}$; basis for H : D_1, D_2 then the basis for H as when $\mathcal{D}_K = K \oplus K$.

$\mathcal{D}_K = K \oplus K$:

γ_1: $D_{u_8(1,3),U_1} + u_8(1,3) \otimes (\varepsilon_2 - \varepsilon_1)$: (1, 1, 0, 0, -1, 2);

γ_2: $D_{u_1(1,2),U_1} - 2U_1(1,2) \otimes (\varepsilon_2 - \varepsilon_1)$: (-2, 1, 0, -1, 2, -1);

γ_3: $(0, 0, 0, E_{16} - E_{38})$: (0, 0, -1, 2, -2, 0);

γ_4: $D_{u_1(1,2),U_1} + 2u_1(1,2) \otimes (\varepsilon_2 - \varepsilon_1)$: (2, -1, 0, -1, 2, -1);

γ_5: $D_{u_8(1,3),U_1} - u_8(1,3) \otimes (\varepsilon_2 - \varepsilon_1)$: (-1, -1, 0, 0, -1, 2);

γ_6: $(0, 0, 0, E_{23} - E_{67})$: (0, 0, 2, -1, 0, 0).

The diagram is

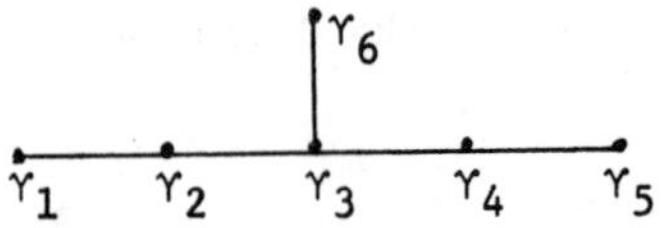

and we have $\gamma_1|_T = \gamma_5|_T = \alpha_4$; $\gamma_2|_T = \gamma_4|_T = \alpha_3$; $\gamma_3|_T = \alpha_2$; $\gamma_6|_T = \alpha_1$. The fundamental weights in Planche V of [3] then are $\pi_1, \ldots, \pi_6$ with

$\pi_i|_T = \lambda_4$ if $i = 1, 5$; $\pi_i|_T = \lambda_3$ if $i = 2, 4$;

$\pi_3|_T = \lambda_2$; $\pi_6|_T = \lambda_1$.

$\mathcal{D} = \mathcal{Q}$:

γ_1: $U_1 \otimes v_2 + 2U_2 \otimes v_2 + D_{v_2,\varepsilon_1}$: (1, 0, -2, 0, 0, 0, 0);

γ_2: $u_8(1,3) \otimes v_1$: (-1, -1, 1, 0, 0, -1, 2);

γ_3: $2U_1 \otimes v_2 + U_2 \otimes v_2 - D_{v_2,\varepsilon_1}$: (1, 2, 0, 0, 0, 0, 0);

γ_4: $u_1(1,2) \otimes v_1$: (-1, 0, -1, 0, -1, 2, -1);

γ_5: $(0, 0, 0, E_{16} - E_{38})$: (0, 0, 0, -1, 2, -2, 0);

γ_6: $(0, 0, 0, E_{23} - E_{67})$: (0, 0, 0, 2, -1, 0, 0);

γ_7: $U_1 \otimes v_2 - U_2 \otimes v_2 + D_{v_2,\varepsilon_1}$: (1, -2, 2, 0, 0, 0, 0).

The diagram is

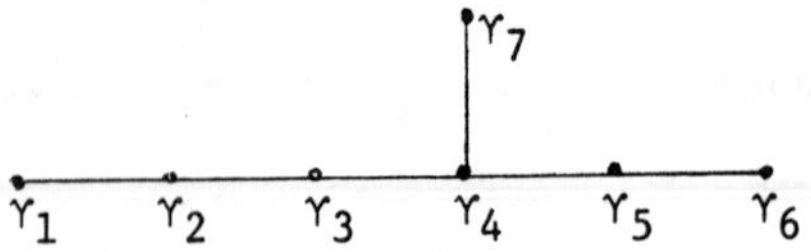

, and

$\gamma_1|_T = 0 = \gamma_3|_T = \gamma_7|_T$; $\gamma_2|_T = \alpha_4$; $\gamma_4|_T = \alpha_3$;

$\gamma_5|_T = \alpha_2$; $\gamma_6|_T = \alpha_1$. The fundamental weights

$\pi_1, \ldots, \pi_7$ of Planche VI of [3] have

$\pi_1|_T = \lambda_4$; $\pi_2|_T = 2\lambda_4$; $\pi_3|_T = \lambda_3 + \lambda_4$;

$\pi_4|_T = 2\lambda_3$; $\pi_5|_T = \lambda_2$; $\pi_6|_T = \lambda_1$: $\pi_7|_T = \lambda_3$.

$\mathcal{D} = \mathcal{O}$:

γ_1: $(0, 0, 0, E_{23} - E_{67})$: (0, 0, 0, 0, 2, -1, 0, 0);

γ_2: $(0, 0, 0, E_{16} - E_{38})$: (0, 0, 0, 0, -1, 2, -2, 0);

γ_3: $D_{u_1(1,2),U_1} + 2u_1(1,2) \otimes (\varepsilon_2 - \varepsilon_1)$: (0, 0, 2, -1, 0, -1, 2, -1);

γ_4: $2U_1 \otimes v_1 + U_2 \otimes v_1 + D_{v_1,\varepsilon_1}$: (-1, 0, -2, 0, 0, 0, 0, 0);

γ_5: The derivation of $\mathcal{O}_K$ annihilating ε_1 and ε_2, sending v_1 to v_2, v_5 to $- v_6$, the remaining basis elements in $E^\perp$ to zero: (2, -1, 0, 0, 0, 0, 0, 0);

γ_6: As for γ_5, but sending v_2 to v_3, v_4 to $- v_5$, the remaining basis elements in $E^\perp$ to zero: (-1, 2, 0, 0, 0, 0, 0, 0);

γ_7: $u_8(1,3) \otimes v_4$: (0, -1, 1, -1, 0, 0, -1, 2);

γ_8: $U_1 \otimes v_1 + 2U_2 \otimes v_1 - D_{v_1,\varepsilon_1}$: (-1, 0, 0, 2, 0, 0, 0, 0).

The diagram is

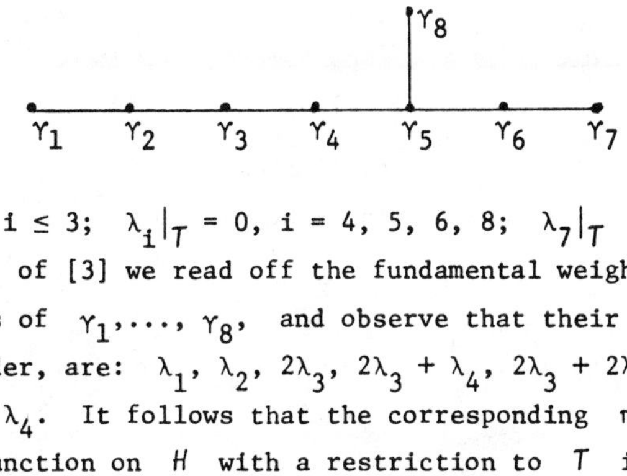

and $\gamma_i|_T = \alpha_i$, $1 \le i \le 3$; $\lambda_i|_T = 0$, $i = 4, 5, 6, 8$; $\lambda_7|_T = \alpha_4$. From Planche VII (p. 268) of [3] we read off the fundamental weights $\pi_1, \ldots, \pi_8$ in terms of $\gamma_1, \ldots, \gamma_8$, and observe that their restrictions to T, in the same order, are: λ_1, λ_2, $2\lambda_3$, $2\lambda_3 + \lambda_4$, $2\lambda_3 + 2\lambda_4$, $\lambda_3 + 2\lambda_4$, $2\lambda_4$, $\lambda_3 + \lambda_4$. It follows that the corresponding π_j is the only dominant integral function on H with a restriction to T in the latter list, except when that restriction is $2\lambda_3 + 2\lambda_4$. In that case $\pi_3 + \pi_7$ and $2\pi_8$, along with π_5, are the only dominant integral functions on H with this restriction to T.

Similar observations on restrictions to T of dominant integral functions on H will be made in the other cases. In all cases, it is only a matter of a little extra care to assure that H is defined over F.

6. $s\ell(3,\mathcal{O})$ AND $sp(6,\mathcal{O})$.

The conventions on $\mathcal{O}_K$ and the notations $\varepsilon_i (i = 1,2)$, $v_i (1 \le i \le 6)$, E, $E^{\perp}$, D_1, D_2 of the last Appendix continue. First, in $s\ell(3,\mathcal{O})_K$, where K is a splitting field for $\mathcal{O}$, we let H_0 be a Cartan subalgebra of $\mathrm{Der}(\mathcal{O})_K$, with basis D_1, D_2; we let $H_1 = \mathrm{diag}\,\{-1, 1, 0\}$, $H_2 = \mathrm{diag}\,\{0, -1, 1\}$ in $s\ell(3,F)$. Then $H = H_0 + H_1 \otimes E + H_2 \otimes E$ is a commutative six-dimensional subalgebra of $s\ell(3,\mathcal{O})_K$ whose adjoint action is diagonalizable and which is its own centralizer. That is, H is a splitting Cartan subalgebra of $L_K = s\ell(3,\mathcal{O})_K$. As in previous Appendices, we abbreviate a full list of root-vectors to a set belonging to a fundamental system; all those for positive roots are obtained by bracketing our set according to the rules for $s\ell(3,\mathcal{O})$. The reader who has carried out that much detail will see at once what the negative root-vectors are.

We follow the scheme of Appendix 5; the final 6-tuple is the set of values of the corresponding γ_i, in order, at the basis for H,

$$(H_1 \otimes \varepsilon_1, H_1 \otimes \varepsilon_2, H_2 \otimes \varepsilon_1, H_2 \otimes \varepsilon_2, D_1, D_2) .$$

γ_1: $E_{12} \otimes \varepsilon_1$: (2, 0, -1, 0, 0, 0);

γ_2: $D_{v_5,\varepsilon_1} + H_1 \otimes v_5 + 2H_2 \otimes v_5$: (-1, 1, 0, 0, -1, 1);

γ_3: $D_{v_4,\varepsilon_1} + H_1 \otimes v_4 + 2H_2 \otimes v_4$: (0, 0, 1, -1, 0, -1);

γ_4: $-D_{v_2,\varepsilon_1} + H_1 \otimes v_2 + 2H_2 \otimes v_2$: (0, 0, -1, 1, 1, -1)

γ_5: $E_{23} \otimes v_5$: (0, -1, 1, 1, -1, 1);

γ_6: That derivation of $\mathcal{O}_K$ annihilating ε_1 and ε_2, sending v_1 to v_3, v_4 to $-v_6$, and the remaining basis elements for $E^{\perp}$ to zero: (0, 0, 0, 0, 1, 1).

The diagram of $\gamma_1, \ldots, \gamma_6$ is

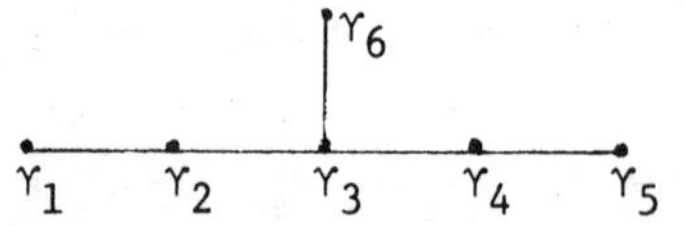

,

with $\gamma_j|_T = 0$ if $j \neq 1, 5$; $\gamma_1|_T = \alpha_1$; $\gamma_5|_T = \alpha_2$. Then one appeals to Planche V (p. 260) of [3] for the fundamental weights $\pi_1, \ldots, \pi_6$, expressed in terms of $\gamma_1, \ldots, \gamma_6$, and finds their restrictions to T, in the same order, to be $2\lambda_1$, $2\lambda_1 + \lambda_2$, $2\lambda_1 + 2\lambda_2$, $\lambda_1 + 2\lambda_2$, $2\lambda_2$, $\lambda_1 + \lambda_2$. The remarks at the end of Appendix 5 apply again here.

When $L = sp(6,\mathcal{O})$, we again denote by $a \mapsto \bar{a}$ the extension of the canonical involution in $\mathcal{O}$ to $\mathcal{O}_K$, where K is a splitting field, and have $\bar{\varepsilon}_1 = \varepsilon_2$. Thus $\varepsilon_1 - \varepsilon_2 \in (\mathcal{O}_0)_K$. With $\mathcal{H}_0$ as above, and with $U_1, U_2 \in \mathcal{M}' \subseteq M_6(F)$ as in §IX.2, $\mathcal{H} = \mathcal{H}_0 + \mathcal{T} \otimes K + K(U_1 \otimes (\varepsilon_2 - \varepsilon_1)) + K(U_2 \otimes (\varepsilon_2 - \varepsilon_1))$ is a Cartan subalgebra of $\mathcal{L}_K$ containing $\mathcal{T}$, of dimension seven, with adjoint action diagonalizable in K. (By starting with reference to a (quadratic) subfield of $\mathcal{O}$, $\mathcal{H}$ may be taken to be defined over F.) Typical root-vectors are displayed below.

To relate bases for $\mathcal{H}$ and for $\mathcal{T}$, it will be convenient to take a basis $h_1, h_2, \ldots, h_5, D_1, D_2$ for $\mathcal{H}$, where $D_1, D_2 \in \mathrm{Der}(\mathcal{O}_K)$ are our basis for $\mathcal{H}_0$, and where $h_1, \ldots, h_5 \in (\mathfrak{sp}(6,F) + (\mathcal{M}' \otimes \mathcal{O}_0))_K$ are as follows:

$$\begin{aligned}
h_1 &= (E_{22} - E_{11}) \otimes \varepsilon_1 - (E_{55} - E_{66}) \otimes \varepsilon_2,\\
h_2 &= (E_{22} - E_{11}) \otimes \varepsilon_2 - (E_{55} - E_{66}) \otimes \varepsilon_1,\\
h_3 &= (E_{33} - E_{22}) \otimes \varepsilon_1 - (E_{44} - E_{55}) \otimes \varepsilon_2,\\
h_4 &= (E_{33} - E_{22}) \otimes \varepsilon_2 - (E_{44} - E_{55}) \otimes \varepsilon_1,\\
h_5 &= H_3.
\end{aligned}$$

Then $H_1 = h_1 + h_2$, $H_2 = h_3 + h_4$, $U_1 \otimes (\varepsilon_1 - \varepsilon_2) = h_1 - h_2$, $U_2 \otimes (\varepsilon_1 - \varepsilon_2) = h_3 - h_4$. The root-vectors relative to $\mathcal{H}$ have the following forms:

i) Root-vectors in $\mathrm{Der}(\mathcal{O}_K)$ annihilating ε_1 and ε_2 --- there are six of these;

ii) Elements $E_{ij} \otimes y - E_{7-j,7-i} \otimes \bar{y}$, y running over our basis for $\mathcal{O}_K$, $1 \le i, j \le 3$, $i \ne j$ -- there are 48 of these;

iii) Elements $E_{i,7-j} \otimes y + E_{j,7-i} \otimes \bar{y}$ and $E_{7-i,j} \otimes y + E_{7-j,i} \otimes \bar{y}$, $1 \le i \le j \le 3$, y as in ii) --- there are 48 of these;

iv) $E_{i,7-i}$ and $E_{7-i,i}$, $1 \le i \le 3$ --- there are six of these;

v) Elements $U_1 \otimes v_i + 2U_2 \otimes v_i - D_{v_i,\varepsilon_1}$; $2U_1 \otimes v_i + U_2 \otimes v_i + D_{v_i,\varepsilon_1}$; $U_1 \otimes v_i - U_2 \otimes v_i - D_{v_i,\varepsilon_1}$, $1 \le i \le 3$ --- there are nine of these;

vi) Elements $U_1 \otimes v_i - U_2 \otimes v_i + D_{v_i,\varepsilon_1}$; $2U_1 \otimes v_i + U_2 \otimes v_i - D_{v_i,\varepsilon_1}$; $U_1 \otimes v_i + 2U_2 \otimes v_i + D_{v_i,\varepsilon_1}$, $4 \le i \le 6$ --- there are nine of these.

Thus 126 root-vectors are obtained. Following the earlier scheme, we list fundamental root-vectors and the sequences of values of the corresponding γ_i at the basis $(h_1, \ldots, h_5, D_1, D_2)$:

γ_1: E_{34}: $(0, 0, -1, -1, 2, 0, 0)$;

γ_2: $E_{23} \otimes v_4 + E_{45} \otimes v_4$: $(0, -1, 1, 1, -1, 1, 0)$;

γ_3: That element of type i) above, sending v_2 to v_3, v_4 to $-v_5$, and annihilating the remaining v_i: $(0, 0, 0, 0, 0, -1, 2)$;

γ_4: $U_1 \otimes v_2 - U_2 \otimes v_2 - D_{v_2, \varepsilon_1}$: $(-1, 1, 1, -1, 0, 1, -1)$;

γ_5: $U_1 \otimes v_6 - U_2 \otimes v_6 + D_{v_6, \varepsilon_1}$: $(1, -1, -1, 1, 0, 1, 0)$;

γ_6: $E_{12} \otimes v_1 + E_{56} \otimes v_1$: $(1, 1, 0, -1, 0, -1, 0)$;

γ_7: $U_1 \otimes v_1 + 2U_2 \otimes v_1 - D_{v_1, \varepsilon_1}$: $(0, 0, -1, 1, 0, -1, 0)$.

The diagram is

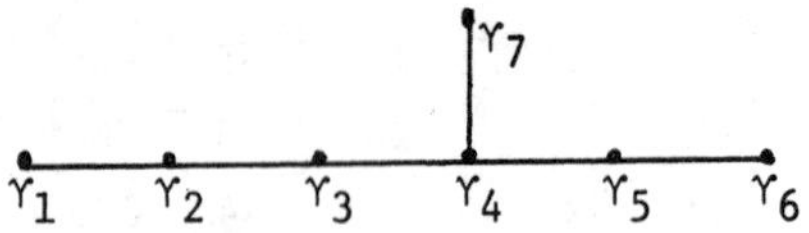

,

and the restriction to T are: $\gamma_6|_T = \alpha_1$, $\gamma_2|_T = \alpha_2$, $\gamma_1|_T = \alpha_3$; $\gamma_i|_T = 0$, otherwise.

The fundamental weights $\pi_1, \ldots, \pi_7$ may be read off from Planche VI, (p. 265) of [3]; using the information above, their respective restrictions to T are:

λ_3, $2\lambda_2$, $\lambda_1 + 2\lambda_2$, $2\lambda_1 + 2\lambda_2$, $2\lambda_1 + \lambda_2$, $2\lambda_1$, $\lambda_1 + \lambda_2$.

All the information used in §IX.3 now follows.

7. $\mathfrak{sl}(2,J)$.

The algebra $L = \mathfrak{sl}(2,J)$ is that of Chapter X, where J is an exceptional central Jordan division algebra. Let K be an extension field of F such that J_K is the split exceptional Jordan algebra of Chapters VIII, IX over K. Let the notations in J_K be those of the "J" of these chapters. Then $\mathrm{Der}(J_K)$ has a splitting Cartan subalgebra with basis H_1, H_2, H_3, H_4, annihilating three orthogonal primitive idempotents e_1, e_2, e_3 in J_K. With $H = E_{22} - E_{11} \in \mathfrak{sl}(2,F)$, the elements

$$H \otimes e_1,\ H \otimes e_2,\ H \otimes e_3,\ H_1,\ H_2,\ H_3,\ H_4$$

freely span a commutative subalgebra of L_K, acting diagonally in the adjoint representation of L_K, and equal to its own centralizer. They therefore span a Cartan subalgebra H of L_K, which may be taken to be defined over F by replacing the span of e_1, e_2, e_3 by a cubic subfield of J. Relative to H, there are in L_K:

48 root-vectors of the form $E_{\ell m} \otimes u_i(j,k)$,

$1 \leq i \leq 8$; $1 \leq j < k \leq 3$; $(\ell,m) = (1,2)$ or $(2,1)$;

24 root-vectors, the root-vectors in $Der(J_K)$ relative to

$\sum_{i=1}^{4} KH_i$ that annihilate e_1, e_2, e_3;

48 more root-vectors, 16 in each of the following sets:

$\{2D_{u_i(1,2),e_2-e_1} \pm H \otimes u_i(1,2),\ 1 \leq i \leq 8\}$;

$\{2D_{u_i(1,3),e_3-e_1} \pm H \otimes u_i(1,3),\ 1 \leq i \leq 8\}$;

$\{2D_{u_i(2,3),e_3-e_2} \pm H \otimes u_i(2,3),\ 1 \leq i \leq 8\}$;

6 final root-vectors $E_{12} \otimes e_i$, $E_{21} \otimes e_i$, $1 \leq i \leq 3$.

These 126 root-vectors, together with $\mathcal{H}$, span L_K.

We designate by $\gamma_1, \ldots, \gamma_7$ the roots to which the elements below belong, and follow each description by the sequence of values of the corresponding γ at $H \otimes e_1$, $H \otimes e_2$, $H \otimes e_3$, H_1, H_2, H_3, H_4:

γ_1: $E_{12} \otimes e_1$; (2, 0, 0, 0, 0, 0, 0).

γ_2: $2D_{u_1(1,2),e_2-e_1} + H \otimes u_1(1,2)$; (-1, 1, 0, -1, 0, 0, 0).

γ_3: The derivation of J_K annihilating e_1, e_2, e_3 and designated in the notation of [24] by $(0, 0, 0, E_{12} - E_{78})$; (0, 0, 0, 2, -1, 0, 0).

γ_4: As for γ_3, by $(0, 0, 0, E_{23} - E_{67})$; (0, 0, 0, -1, 2, -1, -1).

γ_5: As for γ_3, by $(0, 0, 0, E_{35} - E_{46})$; (0, 0, 0, 0, -1, 0, 2).

γ_6: $2D_{u_5(2,3),e_3-e_2} + H \otimes u_5(2,3)$; (0, -1, 1, 0, 0, 0, -1).

γ_7: As for γ_3, by $(0, 0, 0, E_{34} - E_{56})$; (0, 0, 0, 0, -1, 2, 0).

Then $\gamma_1, \ldots, \gamma_7$ are a fundamental system of roots for L_K relative to $\mathcal{H}$, of type E_7:

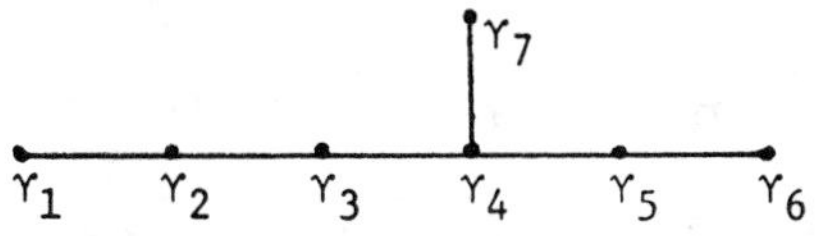

their restrictions $\bar{\gamma}_i$ to T satisfy $\bar{\gamma}_1 = \alpha_1$, $\bar{\gamma}_i = 0$ if $i > 1$.

When the fundamental positive weights $\pi_1, \ldots, \pi_7$ relative to $\gamma_1, \ldots, \gamma_7$ are expressed in terms of the γ_i, one finds (cf. [3], Planche VI, p. 264) that the coefficients of γ_1 are as follows:

In π_6: 1. In π_1 and π_7: $\frac{3}{2}$. In π_2 and π_5: 2. In π_3: $\frac{5}{2}$. In π_4: 3.

On T we have accordingly: $\bar{\pi}_6 = \alpha_1 = 2\lambda_1$; $\bar{\pi}_1 = \bar{\pi}_7 = 3\lambda_1$; $\bar{\pi}_2 = \bar{\pi}_5 = 4\lambda_1$; $\bar{\pi}_3 = 5\lambda_1$; $\bar{\pi}_4 = 6\lambda_1$. Correspondingly we see that the numbers of irreducible representations of L_K whose highest weights have the restrictions below to T are, respectively:

λ_1: none. $2\lambda_1$: one. $3\lambda_1$: two. $4\lambda_1$: three (highest weights $\pi_2, \pi_5, 2\pi_6$). $5\lambda_1$: three ($\pi_3, \pi_1 + \pi_6, \pi_6 + \pi_7$). $6\lambda_1$: seven ($\pi_4, \pi_2 + \pi_6, \pi_5 + \pi_6, 2\pi_1, \pi_1 + \pi_7, 2\pi_7, 3\pi_6$).

8. $G_2(J)$.

We let $L = G_2(J)$, where J is an exceptional Jordan division algebra. The notations are those of Chapter XI. Let the splitting field K be as in Appendix 7, the basis H_1, H_2, H_3, H_4 for a Cartan subalgebra of $\mathrm{Der}(J_K)$ as there, and the basis D_1, D_2 for our maximal split torus T as in §XI.1. In the adjoint representation of L_K, the actions of all the H_i and D_j commute, and they annihilate $(\varepsilon_1 - \varepsilon_2) \otimes (e_2 - e_1)$ and $(\varepsilon_1 - \varepsilon_2) \otimes (e_3 - e_2)$, where the notations are those of §XI.1 and Appendix 7. The last two elements are commuting elements of $(\mathcal{O}_0 \otimes J_0)_K$; together with the H_i and D_j, they span an 8-dimensional commutative subalgebra of L_K. This subalgebra H acts diagonally in the adjoint representation and is its own centralizer, thus is a (splitting) Cartan subalgebra. As in Appendix 7, it may be taken to be defined over F.

The roots of L_K relative to H have root-vectors as follows:

144 of the form $v_r \otimes u_i(j,k)$, $1 \le r \le 6$, $1 \le i \le 8$, $1 \le j < k \le 3$;

24, comprising root-vectors of $\mathrm{Der}(J_K)$ annihilating e_1, e_2, e_3, as for $s\ell(2,J)$;

48, subdivided into the three following sets of 16:

$\{D_{u_i(1,2),e_2-e_1} \pm 2(\varepsilon_1 - \varepsilon_2) \otimes u_i(1,2),\ 1 \le i \le 8\}$

$\{D_{u_i(1,3),e_3-e_1} \pm 2(\varepsilon_1 - \varepsilon_2) \otimes u_i(1,3),\ 1 \le i \le 8\}$

$\{D_{u_i(2,3),e_3-e_2} \pm 2(\varepsilon_1 - \varepsilon_2) \otimes u_i(2,3),\ 1 \le i \le 8\}$;

6, comprising derivations of $\mathcal{O}$ annihilating ε_1 and ε_2, which are already root-vectors of $\mathrm{Der}(\mathcal{O})$ relative to T ;

18 more, described as follows:

$\{v_i \otimes (2e_3 - e_1 - e_2) - D_{v_i,\varepsilon_1},\ 1 \leq i \leq 3\}$

$\{v_i \otimes (-e_1 + 2e_2 - e_3) - D_{v_i,\varepsilon_1}\ 1 \leq i \leq 3\}$

$\{v_i \otimes (2e_1 - e_2 - e_3) - D_{v_i,\varepsilon_1},\ 1 \leq i \leq 3\}$

$\{v_i \otimes (2e_1 - e_2 - e_3) + D_{v_i,\varepsilon_1},\ 4 \leq i \leq 6\}$

$\{v_i \otimes (-e_1 + 2e_2 - e_3) + D_{v_i,\varepsilon_1},\ 4 \leq i \leq 6\}$

$\{v_i \otimes (2e_3 - e_1 - e_2) + D_{v_i,\varepsilon_1},\ 4 \leq i \leq 6\}$.

(See Appendix 5 for the same list in another setting.) There are 240 root-vectors in all.

Designate by $\gamma_1, \ldots, \gamma_8$ the roots to which the elements below belong: each description is followed by the sequence of values of the corresponding γ at the basis D_1, D_2, $(\varepsilon_1 - \varepsilon_2) \otimes (e_2 - \varepsilon_1)$, $(\varepsilon_1 - \varepsilon_2) \otimes (e_3 - e_2)$, H_1, H_2, H_3, H_4 for H :

γ_1: The derivation of $\mathcal{O}$ annihilating ε_1 and ε_2, sending v_1 to v_3, v_4 to $-v_6$ and the other v_i to zero; (1, 1, 0, 0, 0, 0, 0, 0).

γ_2: $v_4 \otimes (2e_2 - e_1 - e_3) + D_{v_4,\varepsilon_1}$; (0, -1, 2, -2, 0, 0, 0, 0).

γ_3: $D_{u_1(1,2),e_2-e_1} - 2(\varepsilon_1 - \varepsilon_2) \otimes u_1(1,2)$; (0, 0, -2, 1, -1, 0, 0, 0).

γ_4: $(0, 0, 0, E_{12} - E_{78}) \in \mathrm{Der}(J_K)$, as denoted under " γ_3" in Appendix 7; (0, 0, 0, 0, 2, -1, 0, 0).

γ_5: $(0, 0, 0, E_{23} - E_{67}) \in \mathrm{Der}(J_K)$; (0, 0, 0, 0, -1, 2, -1, 0).

γ_6: $(0, 0, 0, E_{34} - E_{56}) \in \mathrm{Der}(J_K)$; (0, 0, 0, 0, 0, -1, 2, 0).

γ_7: $D_{u_5(1,3),e_3-e_1} + 2(\varepsilon_1 - \varepsilon_2) \otimes u_5(1,3)$; (0, 0, 1, 1, 0, 0, -1, 0).

γ_8: $(0, 0, 0, E_{35} - E_{46}) \in \mathrm{Der}(J_K)$; (0, 0, 0, 0, 0, -1, 0, 2).

In a tedious exercise one verifies that $\gamma_1, \ldots, \gamma_8$ form a fundamental system of roots for L_K relative to H, of type E_8:

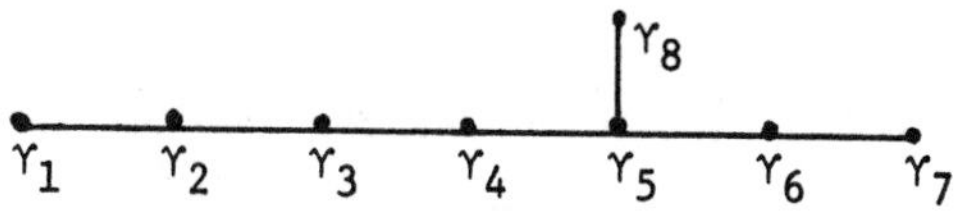

Their values at D_1 and D_2 give their restrictions to T. Thus $\gamma_i|_T = 0$ if $i \neq 1, 2$, while $\gamma_i|_T = \alpha_i$ for $i = 1, 2$.

From [3], Planche VII, we read off the coefficients of γ_1 and γ_2 in the fundamental weights $\pi_1, \ldots, \pi_8$ and use the above to determine the restrictions $\bar{\pi}_i$ to T of the π_i:

$$\bar{\pi}_1 = 2\alpha_1 + 3\alpha_2 = \lambda_1;\ \bar{\pi}_2 = 3\alpha_1 + 6\alpha_2 = \bar{\pi}_8 = 3\lambda_2;$$
$$\bar{\pi}_3 = 4\alpha_1 + 8\alpha_2 = \bar{\pi}_6 = 4\lambda_2;\ \bar{\pi}_4 = 5\alpha_1 + 10\alpha_2 = 5\lambda_2;$$
$$\bar{\pi}_5 = 6\alpha_1 + 12\alpha_2 = 6\lambda_2;\ \bar{\pi}_7 = 2\alpha_1 + 4\alpha_2 = 2\lambda_2.$$

The assertions of §XI.1 based on this appendix are now immediate.

9. $G_2(A)$, FOR SPECIAL CUBIC JORDAN DIVISION ALGEBRAS A.

Both the cases i) and iii) of Chapter XI have the property that over a suitable extension field K, the Jordan algebra A_K becomes the 3 by 3 K-matrices. The split setting is thus the same as that for $\mathfrak{sl}(3,\mathcal{O})$. In $A_K = M_3(K)$, we write e_i for the idempotent matrix-unit E_{ii}, $1 \le i \le 3$, and E_{ij} in general for matrix-units. In L_K, a splitting Cartan subalgebra H has as basis: D_1, D_2, $(\varepsilon_1 - \varepsilon_2) \otimes (e_2 - e_1)$, $(\varepsilon_1 - \varepsilon_2) \otimes (e_3 - e_2)$, $H_1 = \mathrm{ad}(E_{22} - E_{11})$, $H_2 = \mathrm{ad}(E_{33} - E_{22})$, the last two of these in $\mathrm{Der}(A_K)$. (Again, H may be defined over F, starting with a cubic subfield of A.)

The roots of L_K relative to H have the following root-vectors:

36 of the form $v_r \otimes E_{ij}$, $1 \le r \le 6$, $1 \le i, j \le 3$, $i \ne j$;

6, comprising derivations of $\mathcal{O}$ annihilating ε_1 and ε_2, root-vectors of $\mathrm{Der}(\mathcal{O})$ relative to T;

12, consisting of all

$$D_{E_{ij},e_j-e_i} \pm 2(\varepsilon_1 - \varepsilon_2) \otimes E_{ij},\ 1 \le i, j \le 3,\ i \ne j;$$

18 more, described exactly as were the last 18 for $G_2(J)_K$.

In these 72 root-spaces, designate by $\gamma_1, \ldots, \gamma_6$ the roots to which the elements below belong, following each description by the sequence of values of the corresponding γ at the basis D_1, D_2, $(\varepsilon_1 - \varepsilon_2) \otimes (e_2 - e_1)$, $(\varepsilon_1 - \varepsilon_2) \otimes (e_3 - e_2)$, H_1, H_2:

γ_1: $D_{E_{32},e_3-e_2} - 2(\varepsilon_1 - \varepsilon_2) \otimes E_{32}$; (0, 0, -1, 2, 1, -2).

γ_2: $D_{E_{21},e_2-e_1} + 2(\varepsilon_1 - \varepsilon_2) \otimes E_{21}$; (0, 0, 2, -1, -2, 1).

γ_3: $v_4 \otimes E_{12}$; (0, -1, 0, -1, 2, -1).

γ_4: $D_{E_{21},e_2-e_1} - 2(\varepsilon_1 - \varepsilon_2) \otimes E_{21}$; (0, 0, -2, 1, -2, 1).

γ_5: $D_{E_{13},e_3-e_1} + 2(\varepsilon_1 - \varepsilon_2) \otimes E_{13}$; (0, 0, 1, 1, 1, 1).

γ_6: the derivation of $\mathcal{O}$ annihilating ε_1 and ε_2, sending v_1 to v_3, v_4 to $-v_6$, and the other v_i to 0 (as for "γ_1" in $G_2(\mathcal{J})$); (1, 1, 0, 0, 0, 0).

The diagram of the fundamental system $\gamma_1, \ldots, \gamma_6$ for L_K relative to H is

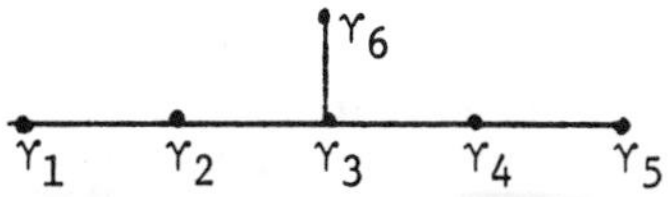

,

and the restrictions $\overline{\gamma}_i$ to T satisfy $\overline{\gamma}_i = 0$ for $i \neq 3, 6$; $\overline{\gamma}_6 = \alpha_1$, $\overline{\gamma}_3 = \alpha_2$. Then Planche V of [3] yields for the corresponding fundamental weights $\gamma_1, \ldots, \gamma_6$ that: $\overline{\pi}_1 = \overline{\pi}_5 = \lambda_2$; $\overline{\pi}_2 = \overline{\pi}_4 = 2\lambda_2$; $\overline{\pi}_3 = 3\lambda_2$; $\overline{\pi}_6 = \lambda_1$. The assertions on i), iii) at the end of §XI.1 are immediate.

Finally, we have the case vi) of §XI.1, where A is a cubic field extension of F, so $\mathrm{Der}(A) = 0$, and $L = \mathrm{Der}(\mathcal{O}) \oplus (\mathcal{O}_0 \otimes A_0)$. The span of D_1, D_2, and $(\varepsilon_1 - \varepsilon_2) \otimes A_0$ is a 4-dimensional Cartan subalgebra H of L containing T. Upon extension to a field K such that $A_K \cong K \times K \times K$, with primitive orthogonal idempotents e_1, e_2, e_3, H_K has as basis D_1, D_2 $(\varepsilon_1 - \varepsilon_2) \otimes (e_2 - e_1)$, $(\varepsilon_1 - \varepsilon_2) \otimes (e_3 - e_2)$, and is a splitting Cartan subalgebra for L_K.

A complete set of 24 root-vectors relative to H_K is given by the description of the last 24 in our list for $G_2(\mathcal{J})_K$ in Appendix 8. Designate by γ_1, γ_2, γ_3, γ_4 roots with root-vectors as indicated below, followed by the sequence of values of the root at the basis D_1, D_2, $(\varepsilon_1 - \varepsilon_2) \otimes (e_2 - e_1)$, $(\varepsilon_1 - \varepsilon_2) \otimes (e_3 - e_2)$ for H_K:

γ_1: $v_4 \otimes (2e_3 - e_1 - e_2) + D_{v_4, \varepsilon_1}$: (0, -1, 0, 2)

γ_2: The derivation of $\mathcal{O}$ annihilating ε_1, ε_2, v_2, v_3, v_5, v_6 and sending v_1 to v_3, v_4 to $-v_6$; (1, 1, 0, 0)

γ_3: $v_4 \otimes (2e_2 - e_1 - e_3) + D_{v_4, \varepsilon_1}$; (0, -1, 2, -2)

γ_4: $v_4 \otimes (2e_1 - e_2 - e_3) + D_{v_4, \varepsilon_1}$: (0, -1, -2, 0).

The diagram of the fundamental system γ_1, γ_2, γ_3, γ_4 is

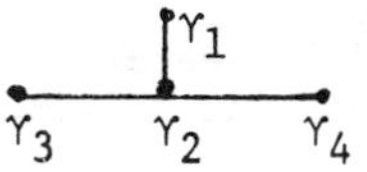

,

with restrictions to T such that $\overline{\gamma}_2 = \alpha_1$, $\overline{\gamma}_i = \alpha_2$ otherwise. For the corresponding fundamental weights $\pi_1, \ldots, \pi_4$ we have $\overline{\pi}_2 = \lambda_1$ and $\overline{\pi}_i = \lambda_2$ for $i \neq 2$. This is the source of the remaining information in §XI.1.

REFERENCES

[1] Albert, A. A., Structure of Algebras. Amer. Math. Soc. Colloq. Publs. Vol. XXIV. Amer. Math. Society, New York, 1939.

[2] Allison, B. N., Isomorphism of simple Lie algebras. Trans. Amer. Math. Soc. 177 (1973), 173-190.

[3] Bourbaki, N., Groupes et algèbres de Lie, Chaps. IV-VI. Éléments de Mathématique, Fasc. XXXIV. Hermann, Paris, 1968

[4] Cartan, E., Les groupes projectifs qui ne laissent invariante aucune multiplicité plane. Bull. Soc. Math. de France 41 (1913), 53-96.

[5] _________, Les groupes projectifs continus réels qui ne laissent invariante aucune multiplicité plane. J. Math. pures et appliquées 10 (1914), 149-186.

[6] Chevalley, C., The Algebraic Theory of Spinors. Columbia University Press, New York, 1954.

[7] Curtis, C. W. and Reiner, I., Representation Theory of Finite Groups and Associative Algebras. Pure and Applied Math., Vol. XI. Interscience, New York, 1962.

[8] Green, J. A., Polynomial Representations of GL_n. Springer Lecture Notes in Mathematics, Vol. 830. Springer Verlag, New York-Heidelberg-Berlin, 1980.

[9] Haile, D., On central simple algebras of given exponent. Jour. of Algebra 57 (1979), 449-465.

[10] Harris, B., Centralizers in Jordan algebras. Pacific J. Math. 8 (1958), 757-790.

[11] Humphreys, J. E., Introduction to Lie algebras and Representation Theory. Graduate Texts in Math., No. 9. Springer, New York, 1972.

[12] Jacobson, N., Basic Algebra II. Freeman, San Francisco, 1980.

[13] ___________, Exceptional Lie Algebras. Lecture Notes in Pure and Applied Math., Vol. I. Marcel Dekker, New York, 1971.

[14] ___________, Lectures in Abstract Algebra. Vol. I, Basic Concepts. Van Nostrand, New York, 1951.

[15] ___________, Lie Algebras. Tracts in Pure and Applied Math., No. 10. Interscience, New York, 1962.

[16] ___________, Structure of alternative and Jordan bimodules. Osaka J. Math. 6 (1954), 1-71.

[17] ___________, Structure and Representations of Jordan Algebras. Amer. Math. Soc. Colloq. Publs. Vol. XXXIX. Amer. Math. Society, Providence, 1968.

[18] ___________, Triality and Lie algebras of type D_4, Rendiconti Palermo (2) 13 (1964), 1-25.

[19] Kemmer, N., The algebra of meson matrices. Proc. Cambridge Phil. Soc. 39 (1943), 189-196.

[20] Malcev, A. I., Fundamentals of Linear Algebra (in Russian). Moscow, 1948.

[21] Redei, L., Algebra. Vol. 1. Pergamon, New York and London, 1967.

[22] Satake, I., On the theory of reductive algebraic groups over a perfect field. Jour. Math. Soc. Japan 5 (1963), 210-235.

[23] Seligman, G. B., On automorphisms of Lie algebras of classical type II. Trans. Amer. Math. Soc. 94 (1960), 452-482.

[24] _______________, Ibid. III. Trans. Amer. Math. Soc. 97 (1960), 286-316.

[25] _______________, Mappings into symmetric powers. Jour. of Algebra 62 (1980), 455-472.

[26] _______________, Rational Methods in Lie Algebras. Lecture Notes in Pure and Applied Math., Vol. 17. Marcel Dekker, New York, 1976.

[27] Tits, J., Représentations linéaires irréductibles d'un groupe réductif sur un corps quelconque. Jour. f. reine angew. Math. 24 (1971), 196-220.

[28] Weyl, H., The Classical Groups. Princeton Math. Series, Vol. 1. Princeton University Press, Princeton, 1939.

DEPARTMENT OF MATHEMATICS
BOX 2155, YALE STATION
NEW HAVEN, CONNECTICUT 06520

ABCDEFGHIJ–AMS–8987654321